Engineering
—— THE ——
Pre-Industrial Age

Engineering
—— THE ——
Pre-Industrial Age

DICK PARRY

AMBERLEY

For Frances and family

First published 2013

Amberley Publishing
The Hill, Stroud
Gloucestershire, GL5 4EP

www.amberley-books.com

British Library Cataloguing in Publication Data.
A catalogue record for this book is available from the British Library.

ISBN 978 1 4456 1445 8
E-book ISBN 978 1 4456 1460 1

Typeset in 10.5pt on 12pt Minion Pro.
Typesetting and Origination by Amberley Publishing.
Printed in the UK.

Contents

Preface

The period embraced by the term 'Pre-Industrial Age', as used here extended from the end of the Ancient World, usually regarded as the fall of Rome, to around the middle of the eighteenth century. The factor which distinguishes the Pre-Industrial Age, as used here, from the Ancient World and the Modern World is power. In the ancient world animal muscles, including human (often slave) muscles, were almost the only source of power exploited by mankind, except for one single, specific application – the grinding of corn. The Greeks and Romans both used the water wheel for this purpose, and Vitruvius gives a clear description of a water mill. Remains can still be seen at Barbegal in the south of France, of a water mill utilising sixteen overshot wheels to produce some 4.5 tonnes of ground corn per day. While animal muscle remained important in the pre-industrial age right through to the middle of the eighteenth century, water wheels became the dominant sources of power for many applications as well as the grinding of corn. These applications included fulling cloth, tanning leather, making paper, crushing olives, sawing timber, raising water, iron forging and activating bellows for iron production. As early as the twelfth century in Europe, 742 Cistercian monasteries in Europe operated water wheels on a commercial basis for a number of these applications. The Domesday Book compiled in 1086 lists 5,624 water mills, overwhelmingly for corn grinding, but some apparently producing blooms of iron, requiring a conversion of the rotary motion of the water wheel to a reciprocal motion for the forge. By 1700 there may have been as many as half a million water wheels in operation throughout Europe. Wind also provided an important source of power, particularly for grinding corn and for land drainage and reclamation. The eighteenth century saw the first harnessing of steam as a major source of power, and ultimately the decline, although certainly not the demise, of water and wind as sources of power. Steam engines had the advantage of mobility, whereas water wheels and windmills were fixed in position; but this advantage was eventually lost with the development in the nineteenth century of water turbines generating electricity, and the consequent construction of major hydro-electric schemes worldwide. Environmental concerns today are leading to consideration and implementation of other possible ways of harnessing water power, such as exploiting undersea tidal currents, and the construction of multitudes of wind turbines.

Other factors introduced in the eighteenth century influencing engineering methods included the cheap mass production of iron, the introduction of mathematics and the principles of physics and mechanics into the design process, and the setting up of major schools of engineering such as the Ecole des Ponts et Chaussées in Paris, all of which transformed the work of civil engineers.

The subject matter of this book is primarily concerned with civil engineering achievements in Europe in the pre-industrial age. Throughout this period, as now, civil engineering works had an overwhelming influence on the lives people led: roads, bridges, tunnels, canals, river improvements, fortifications, cathedrals and other major public buildings, water supply, land drainage and reclamation. Although the term Civil Engineer was not coined until the eighteenth century, and is credited to the British engineer John Smeaton, those responsible for these works throughout the pre-industrial age were certainly civil engineers in the modern sense, although without the analytical and design tools available to the modern professional. Individuals described by such terms as architect, master builder or master mason would often encompass within their expertise and responsibilities functions now undertaken separately and distinctly by architects and civil engineers, and often other specialisations. It is of interest that Procopius, writing in the sixth century, refers to Anthemius of Tralles and Isidorus of Miletus, the builders of the Hagia Sophia in Constantinople, as *mechanikos* or *mechanopoios*, whereas he uses the term *architectôn* to describe Apollodorus of Damascus, the builder of Trajan's bridge across the Danube. The term 'engineer' or 'ingeniator' in the pre-industrial age usually referred to a military engineer engaged in the creation of weapons and fortifications. There were some exceptions, however: for example, Ailnoth (1157–90), who had the title of King's Engineer (Ingeiniator) responsible for various works, including the Tower of London and Westminster Palace.

Very few names of the builders are known for major structures and works built in the ancient world; Apollodorus mentioned above is an exception, as are Imhotep (Step Pyramid, Egypt, approx. 2600 BC), Eupalinos (Samian tunnel, sixth century BC) and Caius Julius Lacer (Alcantara Bridge, Spain, AD104), whose body lies interred close to the end of the bridge. Nothing is known of their other achievements, or of their private lives. The names of some of the builders who created the great works of the pre-industrial age are known, together, in some cases, with their other achievements and even something of their private lives, either through a study of their works or their own reports and writings or the writings of others. Not unusually these achievers were multi-talented. Thus, at least five of the 'Lives' in Vasari's *Artists of the Renaissance* had claims to outstanding engineering achievements, although in the case of Leonardo da Vinci these resided mostly in his head and in innumerable sketches on paper. Important achievements by Giotto, Brunelleschi, Alberti and Michaelangelo can still be seen and are still in use today, Brunelleschi's Dome of the Santa Maria del Fiore having outstanding claims to being one of the greatest structural achievements of all time.

Many of the civil engineering works of the pre-industrial age are still in use today, including major structures such as the great cathedrals and mosques, and bridges taking modern traffic. Many canals, such as the Canal du Midi, are still in use today, both for commercial purposes and leisure activities. The landforms in which many people now work and live, such as the Netherlands or the fenlands of England, were fashioned by the drainage and land reclamation schemes of this period. Thus, it is surely important that we understand how and why, and in what context, these works were created.

1

The Creators

Civil engineering works are created equally by those who design them and those working on site to construct them. In the early or medieval period of the pre-industrial age the designer of a structure would usually have had many years experience on the job, preparing him for the role of Master Builder overseeing major works, and may often have been directly involved in its construction, effectively as an artisan overseer in a hands-on capacity. Many sons followed fathers or uncles into the building industry, often down through a number of generations, and consequently would have been imbued with a knowledge of building from an early age.

Throughout much of the pre-industrial age, skilled masons occupied a niche somewhat apart from most other occupations, resulting from the itinerant nature of their work. Except in very large cities such as Paris (population 200,000), it was not feasible to establish guilds or corporations to protect their interests. The work of a skilled mason could well take him during his lifetime to many parts of France, England and Germany, and perhaps even further afield than this. Medieval masons moved around very much as many civil engineers do today and for the same reasons: their skills were in demand and the major projects on which they worked were one-off enterprises, and when a job was finished the next one requiring their services might be hundreds of miles away. Master masons often worked simultaneously on widely dispersed projects, while aspiring master masons travelled for experience and to learn new techniques. Before the introduction of movable-type printing in Europe around 1452, the more enlightened aspiring master builders, and even experienced masons, accumulated their own store of information on parchment, making sketches of building techniques and various devices they witnessed and appending copious notes. Most of these personal documents have long since perished, but fortunately some of the remarkable sketches and notes recorded by the French master mason Villard de Honnecourt have been preserved. Many of Villard's sketches depicted Gothic building techniques such as flying buttresses and parts of specific structures that won his imagination, including one bay of the nave of Rheims Cathedral, a tower of Laon Cathedral, and the rose window on the west front and the labyrinth (which he sketched back-to-front) of Chartres Cathedral.

Instead of forming guilds, medieval masons established lodges on each building site, a concession won from the bishops in the early days of French Gothic building, these lodges becoming the centre of Masonic power and independence. In the earlier examples the one-storied structures had compartments for a workshop, for storing tools and preparing masonry, and a room where they could eat their meals and meet to swap notes, discuss the job and air grievances. By the middle of the thirteenth century some lodges also contained dormitories, dining halls and a library for storing structural and architectural drawings. Lodge leaders negotiated rates of pay and other site conditions; they took responsibility for ensuring a good standard of workmanship and for the training of apprentices. A mason could travel throughout much of western Europe, leaving his money and possessions behind for safety, needing only the secret password to be accepted into any lodge. He would be provided with food and subsistence, in exchange for which he would describe with the aid of his own sketches developments and technical innovations he had seen on his travels. After making a tour of the site and perhaps adding to his sketches and notes, he set out for the next lodge on his itinerary, possibly even provided with a fresh mount.

The unskilled work of digging, fetching and carrying on site usually fell to labourers recruited from the local population, perhaps farm workers or serfs fleeing into the towns to escape their landlords, or offspring of large families with insufficient income to support them. Sometimes over-zealous or over-pious local people would offer their services voluntarily, which, on some sites at least, brought a strong reaction from the workmen, leading to bloodshed. On the other hand, where labour was short and the workmen's jobs not threatened, voluntary labour was tolerated or even welcomed. In the construction of Chartres Cathedral volunteers from every class of society contributed their labour, even harnessing themselves to carts to drag huge blocks of stone from the quarry, 11 km away. In this instance the workmen may well have been thankful for the assistance of the volunteers.

Working hours on site lasted from dawn to dusk and consequently varied through the year, the working day in the summer extending to as much as 16 hours, with two hours allowed for resting and meals. In addition to Sundays, holidays included upwards of thirty feast days a year and site activity often stopped at noon on the eve of these holidays. Holidays were unpaid, as was time lost through snow, frost or heavy rain. Notwithstanding the long hours, site workers enjoyed a much greater degree of freedom than most other workers at the time. Furthermore, younger members of the workforce, however humble their origins, had prospects of being offered an apprenticeship in one of the skilled occupations. Skilled workers included masons, carpenters, sculptors, stone cutters, stone fitters, mortar mixers, lead workers, plasterers, tilers, smiths and scaffolders. Those with particular talents, some luck and the ability to grasp opportunities as they presented themselves, could rise through the ranks to become foremen, master masons or engineer/architects. There appears to have been little distinction between master masons and architects at the time, both of whom also embraced the role of civil engineer, a term not introduced until the eighteenth century. They were professional men appointed directly by lords, bishops or even kings to design buildings, draw up plans and direct site operations. They could negotiate their own stipends, in some cases upwards of 2 shillings a day, compared to the average 4 pence a day paid to the skilled masons on site. A stipend of 4 pence a day in medieval times would have allowed a married man with two or three children to live with some degree of comfort, but at the lower end of the scale life would not have been easy for a married labourer trying to exist on a half-penny a day.

Throughout the pre-industrial age, there were men who envisaged or instituted great works without having been born into, or previously directly involved in, such works, but who knew how to get any specialist advice they needed. Not unusually, in the medieval period, they came from religious orders, such as Peter Colechurch, responsible for Old London Bridge, having earlier repaired the timber structure that it replaced, and Abbot Suger, who initiated the great period of Gothic construction. In the post-medieval period there emerged in some cases remarkable multi-talented individuals with no basic training in civil engineering projects and structures, who conceived, or were important contributors to, outstanding and complex works. Their names are known, and, in some cases, something about their lives from their own writings or the writings of others. A brief look at the lives of some of those responsible for major achievements in the pre-industrial age reveals a remarkable cross-section of individuals, some with little or no previous related experience, but with an almost messianic drive to succeed.

Abbot Suger (1081–1151)

High groined vaulted ceilings, pointed arches, huge stained glass windows practically bespeak the very spirit of Christianity, so it is appropriate that a senior Church cleric provided the driving force to combine these elements, none of them entirely original, except perhaps in relation to size, to create a new concept in architectural space enclosure. But the man himself was as remarkable as the new architecture – not only one of the most senior Church clerics in France but, for a time, Regent (virtually the first minister) of the country and a formidable historian. And, in addition and not least, a showman. He organised a magnificent ceremony to consecrate the opening of his reconstructed Abbey of St-Denis, inviting the King and all the nobles of the realm, as well as bishops and archbishops from the other major sees, such as Beauvais, Chartres, Rheims, Sens, Senlis and Soissons. They were dazzled; and this new concept in architectural space enclosure took off. In order to ensure that his name would not be forgotten, he had four images of himself placed in the church, together with various inscriptions on stone in his honour.

He was in fact one of two remarkable churchmen at the time making a break from the Cluniac order, but they could hardly have been more different as individuals. Suger's motivation came from his wish to escape the gloom of Cluniac Romanesque architecture, while St Bernard's motivation came from his dismay as he viewed the lapse among Cluniac monks in their observance of the strict principles set up by St Benedict when he first established their order at Monte Cassino in the sixth century. In a letter to the Cluniacs he railed against the immense size of their churches and their taste for expensive luxuries, their coating of church walls and statues in gold, while the poor groaned in hunger and their children went naked. Abbot Suger and St Bernard (1090–1153) were breaking from an establishment with some 1,400 monasteries and dependencies throughout Europe at the start of the twelfth century, centred on the vast abbey church at Cluny itself, which could house up to 400 monks and 2,000 visitors.

St Bernard, a young man of aristocratic birth, together with a number of his friends, joined a small monastery at Cîteaux, and from there he founded the Cistercian order of monks, which, by the time of his death in 1152, had established 343 monasteries, and 530 before the end of the century, throughout Europe. Many were in remote wilderness areas, in which they built up large land holdings and had access to abundant water power, giving them a solid base from which to establish a leading role in the economy of Europe, notably in areas such as agriculture, wine, wool, smelting iron ore and forging iron tools, locks and clamps. Suger remained a Benedictine, and indeed his new architecture was spurred by his wish to create enhanced space and natural light within his structure to be better able to display his gold covered altar and the luxurious treasures thereon, his collection of relics and the lavish decorations, and an enhanced ambulatory area for more people to see them. In correspondence between them Suger must have succeeded in convincing St Bernard that he had moderated his ways, or perhaps the latter took a pragmatic view, and they became friends; or anyway ostensibly so, as befitted two outstanding persons, one close to the papacy and the other the most important political figure in the country.

Little is known of Suger's early life, and he himself never denied coming from humble origins, perhaps even being the son of a serf. He may well have exaggerated the humbleness his origins. It is remarkable for a boy from such lowly circumstances to have been, at the age of eleven, enrolled into the monastic school of Saint-Denis de l'Estree, where the list of students included the sons of nobility, princes of the realm and, not least, the future King of France. The school lay within the shadow of the Abbey and he claimed later in life to have harboured an ambition to rebuild it while still at this school. It sounds a little bit like a good story. In 1106 he became secretary to Abbot Adam of St-Denis after, apparently, two years final preparation at the monastic school of St-Benoît-Sur-Loire near Orléans. He moved on to become provost of Berneval in Normandy in 1107 and of Toury in Beauce in 1109. His boyhood friend, King Louis VI (Louis the Fat), evidently followed his career with some interest. Impressed with Suger's diplomatic and political skills, and needing someone to reinforce political and Church ties, he sent him in 1118 to the Court of Gelasius II at Maguelone in the south of France and subsequently in 1121 to that of Callistus II in Rome.

Although appointed Abbot of St-Denis in 1121, while still in Rome, he remained close to the King as his personal advisor and on his return to France spent much of his time up to 1127 at Court. Louis VI died in 1137, to be succeeded by his oldest surviving son, Louis, then only 17 years of age and neither ready nor suited to be king. Louis VII came to rely as much, or more, than his father on Suger's counsel, which failed him with respect to his marriage to Eleanor of Aquitaine in the same year as his accession. Suger certainly approved the marriage and accompanied the young prince to Aquitaine on the occasion of the nuptials. The marriage ended in divorce in 1152.

Concerned by possible threats to the crown by local nobles, Suger counselled against Louis' departure on the Second Crusade. The King chose on this occasion, unwisely, to ignore his advice and set out for the Holy Land in 1147, encouraged and accompanied by Eleanor, anxious for her husband to throw his support behind her uncle, Raymond of Antioch. Despite suffering losses on the way when ambushed by the Turks, the King and his army reached the Holy Land in 1148, but then took part in a disastrous siege of Damascus. The King returned to France in 1149, no doubt wiser, but his venture had cost his treasury and military dearly. Before setting out, he had hedged his bets by appointing Suger as his Regent to run the country during his absence, which he did with such success that the King gave him the title 'Father of the Country'. Surprisingly, in the event, Suger was equipping an army and preparing to set out for the Holy Land when he died in 1151.

Abbot Suger's important writings include an encomium on the life of Louis VI and a more balanced account of Louis VII, giving historians valuable insights into the lives of these two monarchs. Perhaps even more significantly, for architectural historians at least, he wrote voluminously on his reconstruction of the Abbey of St-Denis, in *Liber de Rebus in Administratione sua Gestis*, *Libellus Alter de Consecratione Ecclesiae Sancti Dionysii* and *Ordinatio*, giving details of the works and treasures of the church.

Villard de Honnecourt (thirteenth century)

What little is known about Villard de Honnecourt has been deduced from the annotations accompanying his many sketches, thought to have been made between 1225 and 1250 and now housed in the Bibliothèque Nationale in Paris. They reveal two certainties: firstly, that he had an insatiable curiosity about the world around him and secondly, that his travels took him to many parts of France, and to Switzerland and Hungary. The thirty-three parchment leaves of his sketches, inscribed both recto and verso, embrace such diverse subject matters as architectural and sculptural features of a number of Gothic cathedrals, various machines, solutions to geometrical and trigonometry problems, animals such as cats, dogs, horses, bears and lions and other wildlife including birds, hares, hedgehogs, dragonflies, bees and snails. His human sketches included the crucifix as well as nudes and figures in various attires, paying particular attention to the folds in the cloth, perhaps for the benefit of cathedral sculptors. He sketched various features of cathedrals at Vaucelles and Cambrai – both close to his birth village of Honnecourt – Chartres, Laon, Meaux and Reims in France and Lausanne in Switzerland, and he mentions visiting Hungary in response to a commission to build a cathedral at Kosice.

Villard's rank within the building fraternity is not known. His notes certainly show that he understood the complexities of Gothic construction, and the extent of his travels to a number of cathedrals may suggest that he enjoyed the status of a master builder or master mason with special skills that were in demand. On the other hand, he may simply have been a man of modest building skills, but with an insatiable curiosity about cathedral construction and everything in the world around him, which motivated both his travel and the need to sketch the things he observed. He saw his cathedral sketches at least as more than just a personal interest, claiming, as he does in his introduction, that his book could be a great help in instructing the principles of masonry and carpentry and that it contained methods of portraiture and line drawing as dictated by the laws of geometry. His perceived educational purpose is strongly supported by one particular page of many sketches illustrating, among other things, how to: measure the diameter of a partly visible column; find the mid-point of a drawn circle; cut the mould and the

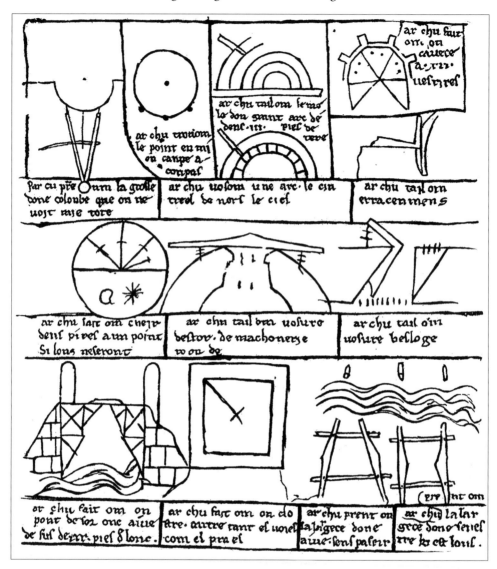

Part of a page of sketches by Villard de Honnecourt showing *Top*: how to: measure the diameter of a partly visible column; find the mid-point of a drawn circle; cut the mould of a three-foot arch; arch a vault; create an apse with twelve windows; cut the springing stone for an arch. *Middle*: how to: match two nearby stones; cut a voussoir for a round building; cut an oblique voussoir. *Bottom*: how to: make a bridge with 20 foot timbers; lay out a cloister and courtyard; remotely measure the widths of a watercourse and a distant window.

Sketch by Villard de Honnecourt showing a hand-operated mechanism for sawing off piles under water.

springing stone of an arch; cut a regular voussoir and an oblique voussoir; make a bridge over water with 6m long timbers; lay out a cloister with its galleries and courtyard; place the four cornerstones of a cloister without plumbline or level; measure the width of a water course without crossing it; measure the width of a distant window.

His first employment as a young lad may have been at the Cistercian monastery of Vaucelles, long since demolished, but his sketch of the choir ground plan, perhaps his earliest meaningful sketch, has been confirmed as accurate by archaeologists. He may have worked next at nearby Cambrai, his sketches of which seem to have been a direct copy of architectural drawings. His further architectural sketches included a double row of flying buttresses; elevations of Reims and accompanying text describing various features, including its internal passageways to allow circulation in case of fire; towers of Laon and Meaux cathedrals; and the rose window and labyrinth at Chartres. Curiously, he drew the Chartres labyrinth in mirror image, and in the case of the Reims elevations and the Laon tower 'updated' them to look apparently as he would have constructed them.

Geometric representations abound in his sketches, including efforts to reduce animal and human images to combinations of triangular forms or, in some cases, triangular and rectangular forms. Clearly fascinated by the mechanical arts, he depicted various machines and gadgetry, possibly some of his own devising, including a hand operated arrangement for sawing off piles under water, a water-powered device for sawing planks, and a weight-driven clockwork mechanism driving a spindle connecting to the statue of an angel on the roof of a great church, causing it to rotate once every twenty-four hours, keeping a finger pointing towards the sun. He also turned his mind to military engineering, depicting and describing a powerful catapult requiring two windlasses to haul down the launching pole and a crossbow fitted with a pinhole sight so that it 'never misses'.

Henry Yevele (*c.* 1320–1400)

Born around 1320 near Yeavely in Derbyshire, Henry Yevele seems to have been fortunate enough at the time of the Black Death in 1348/9 to have been still living and working in an area with death rates much lower than those in the southern counties. Like many craftsmen in the Midlands and more northern counties, he made his way to London after the plague, attracted by the opportunities arising as a result of the high death toll among London masons and an upsurge of work in the capital. Quickly establishing himself, his career really took off after his appointment in 1357 as mason to the Black Prince, son of King Edward III, although initially engaged on fairly minor works.

It is likely that Henry's father Roger practised as a mason too, and Henry's brother Robert certainly did. Henry may have spent some of his early years working on Lichfield Cathedral, the ecclesiastical centre of the region. He also acquired skills in tomb and monument carving in these early years, based on industries associated with the alabaster quarries at nearby Tutbury. His later works included the tombs in Westminster Abbey of Edward III, who died in 1377, and Richard II, who died in 1399, having ordered the construction of his tomb a few years before although still only in his thirties.

Around 1360 the King appointed him 'Dispenser of the King's Works', with particular immediate responsibilities for two major construction works, the Palace of Westminster and the Tower of London, but in effect he became chief architect/engineer for the whole of southern England. Over the next four decades until his death in 1400 his work load encompassed major civic and ecclesiastical buildings, fortifications, bridges, wharves and monumental works. Major Church reconstruction works included the west cloister and new nave for Westminster Abbey in London and extensive work on Canterbury Cathedral, including transformation of the crypt into a chantry chapel, side-walls for the aisles and the interior of the nave. Although the iconic status enjoyed by Westminster Hall in London derives largely from master carpenter Hugh Herland's great hammer beam roof, the masonry work on the structure by Yevele also deserves special mention. At the insistence of Richard II, who loved the Arts and architecture, the reconstructed building replacing the huge old hall, 238 feet long and 69 feet wide and dating from the time of William Rufus around 1098, had to be to be the most beautiful and up-to-date building in the country. Yevele's contribution to the new design included removing internal arcading and two rows of great posts supporting the roof, facing the existing walls and providing a new range of traceried windows, and strengthening the east and west long walls with flying buttresses to take the thrusts from the roof. Preparations for the work had barely started when the King's twenty-seven-year-old wife Anne died in 1394, driving him into the depths of despair and delaying for several months any work on the project.

Yevele's considerable forays into military work included, in Kent, the west gate of Canterbury and the coastal defence castles of Queenborough and Cowling. Built as a garrison fortress only on flat land bordering the mouth of the Medway, his simple but imaginative solution for Queenborough consisted of an outer wall 165 feet in radius with a surrounding moat, and a higher concentric inner wall 66 feet in radius, incorporating a range of buildings and having six semicircular towers attached. With the outer wall entrance with its two round towers situated at the west end of the fortress and the inner gate at the east end, any enemy managing to penetrate the outer gate still had to circumnavigate halfway around the fortress, between the inner and outer walls, under withering fire from both, before reaching the inner gate.

Yevele's work brought him into close contact with Chaucer in the latter's role as Clerk of the King's Works, a position he held from 1389 to 1391, having spent his working life in royal and government service, including controller of customs on wool in the port of London between 1374 and 1386. Despite these responsibilities he produced a substantial body of poems, such as *Troilus and Criseyde*, and English translations in the 1370s and early 1380s, before writing *The Canterbury Tales* in the late 1380s and early 1390s. As one of his first acts as Clerk of the King's Works, Chaucer arranged to pay Yevele the sum of £6 17s 1d in salary arrears, which had accumulated over a period of time. Yevele then resumed work on the repair of a wharf serving the Tower of London.

Chaucer's pilgrims on their way to Canterbury, crossing the Medway on the old timber bridge in 1388, would have seen under construction some 120 feet upstream Yevele's great stone bridge, the second most important bridge in England after London Bridge. Built between 1383 and 1392, the structure of ten arches and a drawbridge span had a length of

566 feet and a width of 14 feet. The financing of the bridge came from local landowner Sir Robert Knolles, one of Edward III's generals, recently returned with spoils from the wars in France. His friend Sir John Cobham financed the building of a chapel at the east end of the bridge. The bridge stood for 475 years until it was demolished and replaced with a cast iron structure in 1857, the *Illustrated London News* publishing a graphic picture showing a large crowd witnessing the blowing up of the old structure, with huge blocks of masonry being thrown up into the air. Yevele's other bridge commissions included a modest structure linking Chelmsford and Moulsham in Essex, but more importantly from 1368 he acted for extensive periods for the remainder of his life as a Warden of London Bridge, conveniently living in a house situated near the north end of the bridge, opposite the church of St Magnus. This period saw major reconstruction of the bridge, probably in no small measure due to his influence, and the building of a new chapel under his supervision in the Perpendicular Gothic style with pointed arches and fan vaulting.

He became a wealthy man, not primarily from his professional fees, but from entrepreneurial activities, including supplying materials such as masonry, tiles, bricks and plaster of Paris to the building industry, some for buildings for which he himself was the commissioned designer and supervisor. He had an interest in the Purbeck marble quarries. He also bought and sold many properties in London, Essex and Kent, and on his death bequeathed to his second wife, Katherine, many tenements and houses, watermills, two quays and a brewery called La Glene, lands at Deptford and Greenwich, and an estate near Purfleet in Essex where he spent his declining years.

He died on 21 August 1400 and was buried in the church of St Magnus, in a tomb of his own design. Chaucer died a few weeks later and was entombed in Westminster Abbey.

Filippo Brunelleschi (1377–1446)

Vasari describes Brunelleschi as having been of unprepossessing appearance, but says that such men are often very magnanimous and pure in heart, and when nobility is added to their other qualities, they may confidently be expected to work miracles. His greatest work, the Duomo of the Cathedral of Santa Maria del Fiore must surely have seemed like something of a miracle to many on its completion: suffice it to say that it rates as one of the greatest structural engineering achievements of all time. Like many sons before and since, he disappointed his father by not following in his profession as a notary; but his clearly enlightened father, noting the boy's obvious obsession with sculpting and mechanical devices, set him to learn arithmetic and writing and apprenticed him to a goldsmith to study design. He became a master goldsmith in 1398, giving him direct access to important figures in the city and a solid basis to fall back on if necessary while he pursued his other interests, including sculpting and perspective and assisting on a number of building projects. It also led to his becoming a member of the Consiglio del Popolo, one of two bodies comprising 500 members representing the people of Florence with no specific legislative powers, but which could vote on laws proposed by the governing body, the Signoria.

In sculpting, he no doubt knew that his talents did not match those of his close friend and companion Donatello, who, although ten years younger, had had the advantage of working in Ghiberti's studio. In 1401 Brunelleschi and Donatello both submitted proposed scenes for the two bronze doors of the church and baptistry of San Giovanni, in competition with other artists, and according to Vasari, when the commission was awarded to Ghiberti they took off together to Rome to spend several years there, Brunelleschi to study architecture and Donatello to study sculpture. In fact they seem mostly to have worked together for a while studying and measuring ancient buildings and ruins, but Donatello returned to Florence, leaving his friend Filippo busily making detailed drawings of temples, basilicas, aqueducts, baths and amphitheatres and the methods used in their construction, such as the dovetailing and binding together of stones. He also noted the presence of holes in the stone blocks that took the iron grips used in hauling them up, a technique he later adopted himself.

Some doubt is cast by Eugenio Battista on Vasari's account, as it came from uncertain sources, and as he makes clear that little is known about Brunelleschi's activities before 1418, when he was already forty years of age. He surmises that Brunelleschi may well have visited Rome before his submission for the baptistry doors in 1401, and that he made another visit before 1417, which isn't in conflict with Vasari, and a third visit at the start of the 1430s. A good deal is known about him and his activities after 1418 and up to his death in 1446.

In 1420 the Opera (Board of Works) of Florence Cathedral, after years of deliberating on how the dome could be built, invited specific proposals and models from a number of masters. Vasari tells the story that Brunelleschi suggested that the commission should go to the person who could successfully stand an egg on its end on a flat piece of marble. None succeeded until Brunelleschi cracked its bottom on the marble, enabling it to stand upright. When the others objected, he retorted that they could also have built the dome if they had seen his model. In the event they entrusted the construction of the dome jointly to Brunelleschi and Ghiberti, largely because Brunelleschi had convinced them it could be built without centring, which would have been prohibitively expensive. Brunelleschi's acceptance of this joint appointment testifies to his eagerness to construct the dome, and probably to a belief that he could rid himself of the unwanted appendage of Ghiberti, which he eventually did. The inclusion of Ghiberti would, in fact, have seemed sensible to most concerned people at the time as Brunelleschi had had little or no experience in major construction works.

His first major commission, the Ospedale degli Innocenti or Foundling Hospital in Florence, came only in 1419, when he was already 42 years of age, notable particularly for its impressive 71 metre long arcade of arches 8 metres high and 5 metres wide. The Opera of the Founding Hospital paid him off in 1427 before its completion, perhaps because other commitments prevented him giving his full attention to the work. Other principal commissions or commitments, other than the Duomo, included the Basilica and Old Sacristy of San Lorenzo, Palazzo di Parte Guelfa, Santo Spirito di Firenze and the Pazzi Chapel. Other civil works planned or undertaken included residential layouts, river dams and canalisation, drainage systems and aqueducts.

Brunelleschi's talents extended to mechanical as well as structural engineering, enabling him to design hoists for lifting heavy building materials, with changeable gears allowing rapid reversal of direction or change of speed. These predated by more than fifty years Leonardo's sketches of mechanical devices. In 1421 the City of Florence gave him a patent for a large transport ship he designed to bring building materials to the site, equipped with cranes to handle heavy cargo. It was the first known patent to be granted.

Although he reputedly despised war and the military, he spent substantial portions of his time between 1423 and 1432 absent from Florence, particularly in Pisa, Pistoia and Milan, designing or advising on military machines and fortifications, largely at the instigation of, and paid for by, the Florentine Opera del Duomo. He devised military machines for transportation, hoisting, scaling walls and attacking towers, including the construction of a huge revolving crane of wood, counterbalanced with blocks of stone, able to lift to high levels heavy cannon and stone blocks and position them accurately.

Leon Battista Alberti (1404–72)

Jacob Burckhardt, in his book *The Civilization of the Renaissance in Italy*, first published in 1860, which initiated modern studies in the achievements of the Renaissance and its leading figures, says of Leon Battista Alberti that in his youth he could, with his feet together, spring over a man's head; also that if he threw a coin in the air it could be heard to ring against a distant roof, and that the wildest horses trembled under him. These talents seem to have been introduced by Burckhardt to support his contention that Alberti had special claims to distinction from a young age. Whatever the truth of these youthful talents, he certainly demonstrated outstanding scholarly and practical talents in his adult life.

He came from a prominent Florentine family, with wealth deriving from commerce and banking, bolstered by having family links throughout Europe and the Near East, their numbers including not just acute businessmen, but also scholars, poets and mathematicians. At the time of his birth in Genoa in 1404, the family had been exiled from Florence for 17 years for political reasons by the republican government, an exile that lasted another 23 years. With their widespread contacts and trading network, it made little difference to their commercial activities and prosperity. Two years after his birth his mother, a widow, died in an outbreak of plague and his father Lorenzo moved firstly with his two illegitimate, but apparently much loved, sons away from the plague to Venice and in 1415 moved on to Padua, now married to a Florentine woman. Lorenzo enrolled Leon Battista as a boarder at the Gymnasium in Padua, the top school in northern Italy, where he became immersed in Latin and the writings of the ancient classicists. From there he moved on to Bologna University for seven years to study law, but found time to pursue his many true interests in the visual arts, in mathematics and science, and in the ancient classical texts. During this time first his father, then his guardian uncle Ricciardo, died, leading to a decline in the family fortunes, and Leon Battista suffered a period of ill-health. He graduated from Bologna in 1428, the same year that Pope Martin V issued a papal bull allowing the Alberti family members to return to Florence.

Between 1430 and 1432 he travelled to France, Belgium and Germany, possibly accompanying Cardinal Albergati, Bishop of Bologna, as his secretary, following which he took up a position in Rome as a Papal Secretary. This allowed him adequate time to pursue his many other interests in ancient classical texts, arts and sculpture, architecture and engineering, and literature and writing. The last included an eclectic mix of short elegies and pastoral poems, a treatise on domestic life in four books and even a funeral oration on his dog. This short stay in Rome lasted only until 1434, when the new Pope, Eugenius, took flight to Florence with his retinue, including Alberti, and where he and the Papacy spent the next nine years. He then returned to Rome to live there for the rest of his life.

In his first period in Rome, his interests appear to have been mainly in painting, sculpture and the study of perspective, a subject that engaged the minds of a number of leading minds of the quattrocento. It also led to a close friendship between Alberti and Brunelleschi, despite a generation's difference in their age. Alberti returned to Florence at the time when the dome of Florence Cathedral was nearing completion, a construction of which he wrote, 'Surely a feat of engineering, if I am not mistaken, that people did not believe possible these days and was probably equally unknown and unimaginable among the ancients' – praise indeed from a man whose promotion of the architecture of the ancient world played a major part in the 'renaissance' itself. Their shared interests and experimentation in perspective drew the two men together, leading to Alberti dedicating to Brunelleschi his 1436 treatise *On Painting*, in which he deals at length with the mechanics of perspective.

Other than a bronze self-portrait, nothing positively identifiable survives of Alberti's paintings or sculpture, but several of his buildings do and, most importantly, *de Re Aedificatoria*, his *Ten Books of Architecture*. Buildings wholly or in part designed by him for wealthy clients can still be seen and appreciated. In the unfinished *Tempio Malatestiano* in Rimini, he adopted the novel plan of encapsulating the existing medieval brick church of San Francesco in a classical marble shell, the arched façade joining to side arcades, with the openings positioned to allow light through to the windows of the original chapels. In something of the same spirit, in 1457 Alberti designed for his wealthy Florentine client Giovanni di Paolo Rucellai a new façade for the late thirteenth-century church of Santa Maria Novella, but in this case had to incorporate into his design an existing lowest portion of three doorways and six slightly pointed arched tombs built a century earlier. The continued use of pink stone with black and dark green marble inlays gave a seamless merging of the lowest portion with Alberti's classical upper portion, with its rounded arches, circular central window and scrolls and inlaid discs and prominent pediment. The proportions are such that the façade fits exactly into a square, the lower half having the full width of the square and the pediment of the upper part the central half width. Alberti also designed for this client in Florence a palazzo and loggio. Towards the end of his career he designed two churches in Mantua for Ludovico Gonzaga, both of which were completed after his death.

His intent in writing *de Re Aedificatoria* he states quite clearly himself:

... we shall collect and transcribe into this our Work, all the most curious and useful Observations left us by the Ancients, and which they gathered in the actual Execution

In his fifteenth-century façade for the thirteenth-century church of the Santa Maria Novella in Florence, Alberti successfully incorporated into his classical design an existing lowest portion of three doorways and six slightly pointed arched tombs.

of their Works; and to these we shall join whatever we ourselves may have discovered by our Study, Application and Labour, that seems likely to be of use.

In other words, he is in fact updating and expanding Vitruvius' *de Architectura* to produce a comprehensive handbook for fifteenth-century engineers and architects. Although completed in manuscript in 1452 it was only published, by his brother, in 1485, thirteen years after his death. The first English translation from the original Latin appeared in 1726, attesting to its perceived importance at that time.

The text covers a wide range of topics from the general to the specific, from matters concerned with selecting suitable sites for buildings or even whole cities to methods of moving and raising large stones, from ensuring that buildings conform to classical proportions to ridding buildings of pests and vermin, described as 'Inconveniences'. For example, he advises readers that Solinus claimed that serpents could be driven away by 'strewing a Place with some of the dust from the Isle of Thanet in Britain'. A wide range of buildings come within its remit – public buildings, fortifications, hospitals, schools, sepulchres, bridges and even the first known reference in Europe to the pound lock. Advice is given on building materials – stone, brick, timber and mortar – and on construction equipment. Water is dealt with – locating it, conveying and using it and the provision of drains and sewers. Where ground conditions at a site are variable, he recommends avoiding the softer areas by linking columns by turned (inverted) arch foundations to concentrate foundation loads onto the firmer areas.

Francesco di Giorgio Martini (1439–1502)

In 1490 Francesco di Giorgio participated in a meeting of engineers and architects called to Milan to exchange views on completing the cathedral in that city, and while there met the 38-year-old Leonardo da Vinci, working in the service of the Sforza family. It was a meeting between the ultimate practical and pragmatic engineer, with many completed works to his credit, and the ultimate intellectual with a mind so engorged with new ideas and thoughts that he had difficulty applying himself to the completion of anything. Nevertheless, the two men seem to have struck up an instant rapport, to the extent that Francesco asked Leonardo to accompany him to Parma, where he was being consulted on the construction of a cathedral.

Born in Siena in 1439, Francesco di Giorgio came from a modest background, and studied painting and sculpture in his early life, the latter skill staying with him, although he primarily regarded himself as an engineer and architect. In 1469 the Siennese city authorities entrusted him with a responsible role in maintaining the water-supply, fountains and aqueducts for the city. His fountain in the main square still exists. His most productive period came, however, after accepting in 1477 an invitation by Federigo Montefeltre, Duke of Urbino, to be military engineer and architect for that city. A man of letters and a patron of the arts, Federigo had created a city with a well-stocked library and where high priority attended scientific, technical and intellectual pursuits. Francesco undertook for Federigo, and subsequently for his son Guidobaldo, the construction, or at least a substantial part of the construction, of the palace of Urbino and contributed as well to work of the ducal palace of Gubbio. His fortifications included the construction of several hill-top fortresses to protect the approaches to Urbino. He continued supervising this work after returning to Siena in 1486 and also took on other commissions in other parts of the country, including fortifications in Casole d'Elsa in 1487 and Lucignano in 1490. He prepared, too, detailed plans for the façade of Siena Cathedral.

During his time at Urbino, and possibly for a period before going there, he worked on his three-part treatise, *Trattato dell' Architettur civile e militare*, completed sometime around 1480, and the first known important work on fortification. Incomplete parts of the manuscript are found in various libraries in Italy, and a portion of particular interest is the Ashburnam manuscript in the Laurenziana in Florence, which has marginal annotations by Leonardo. His copiously illustrated treatise embraced architecture (which owed much to Vitruvius), fortifications and machines, the last comprising both a compilation of machines in common use, together with his own inventions and ideas, and which also owed much to the publications of Taccola (1382–1453), his Siennese predecessor. Leonardo da Vinci, in turn, derived many of his ideas for machines from Francesco's treatises (and probably directly from Taccola as well) and marginal notes by Leonardo in Francesco's manuscripts led to a long-held belief that he was the author.

Francesco's writings on architecture included site selection and planning, such as having nearby sources of drinking water, avoiding low-lying ground or either ore-bearing or bituminous ground. He repeats Vitruvius in claiming that land on which sheep prospered was good, healthy land on which to build. On city layout, he recommended a concentric octagonal plan with a central octagonal square and eight roads leading from it to the gates into the city.

The fortress most completely incorporating the ideas expressed by di Giorgio in his treatise was built at Ostia in the 1480s, not by di Giorgio but by Giuliano da San Gallo. In 1478 the two men had become acquainted when Giulliano della Rovere, accompanied by Francesco di Giorgio and the Duke of Urbino, laid siege to the town of Castellina, by order of Pope Sixtus IV. The Florentine Lorenzo de Medici had entrusted the defence of the town to Giuliano da San Gallo, but unable to hold out, he capitulated. And so the two men met and San Gallo no doubt listened carefully to any tips his counterpart may have been disposed to pass on.

Francesco depicts in his treatise many machines, most of them powered by overshot water-wheels, with the water led directly on to receptacle-like vanes from a tapering conduit, rather than in free fall. He also shows horizontal wheels, in effect turbines, driven by a jet of water. His machines feature various gearings and mechanisms, including toothed wheels and lantern wheels, connecting rods and cranks, and screw and rack systems. His depiction of a ball governor predates that of Watt by three centuries. His interest in public works is attested to by a simple dredger and pile-driving equipment with an ingenious release mechanism.

He died in Siena in 1502

Sangallo Brothers – Giuliano (1443–1516) and Antonio (1455–1534)

Two prominent military engineers of the fifteenth and sixteenth centuries were Giuliano da Sangallo and his brother Antonio da Sangallo, both born in Florence, the sons of Francesco Giamberti, a carpenter and architect in the service of Cosimo de' Medici. In his early life Giuliano worked in Florence for Cosimo's son Lorenzo, who commissioned him to build a monastery outside the Florentine gate of San Gallo, after which the brothers changed the family name to Sangallo. In 1478, entrusted by the Florentines with the defence of the town of Castellina, Giuliano found himself opposed to Francesco di Giorgio, who accompanied the Duke of Urbino and his forces who were besieging the town on behalf of Pope Sixtus IV. Giuliano had to agree to an honourable capitulation, leading to probably his first meeting with Francesco.

Giuliano returned to Florence in 1485, spending his time on both fortress design and restoration, such as that at Perugia, and the construction of military machinery, mostly copied from Francesco di Giorgio. He took part in a number of sieges, notably at Pisa in 1509 and Mirandola in 1511. His expertise with machines gained for him, with his brother, the task in 1504 of moving the huge weight of Michelangelo's marble David a considerable distance, from the workshop in Santa Maria del Fiore to where it stands today. Their success in accomplishing this task using capstan and endless screw devices drew the applause of the large watching crowd.

Antonio Sangallo first worked as an understudy to his brother before taking on his own commissions. In some of his early work, he raised the walls of Castel Sant' Angelo in Rome in 1493, then under threat from the French army, and over thirty years later modernised this fortress by adding triangular bastions which can still be seen. His many other works included restoration of Civitavecchia, a pentagonal fortress with bastions and projecting towers. He was still working at the time of his death on the citadel of Del Basso, introducing all the modern features introduced by the Sangallos. The military work of the family, however, did not end there as it was carried on by a nephew of the brothers, also called Antonio Sangallo.

Leonardo da Vinci (1452–1519)

It might be argued that Leonardo da Vinci should not be included in a list of outstanding creators of engineering works as he is not known to have been responsible for any major constructions, although there is evidence that his advice was sought on a number of important civil and military projects. His understanding of such works is well attested to in his many sketches and notes, which also provide an insight for engineers and others today into the minds of those who created the great works of his time, as he copied many of his ideas from his contacts with, or the writings of, others of his time, although invariably improving upon, or greatly extending, their ideas.

Leonardo himself apparently entertained no doubts about his engineering talents, as evidenced in a letter sent by him in 1481 to Ludovico Sforza, the ruler of Milan, written for the entirely pragmatic purpose of seeking a job in a time of shifting conflicts and alliances between city states themselves and between city states, the Papacy, the King of France and the Holy Roman Emperor. He made the following claims:

1. I have a process for the construction of very light bridges, capable of easy transport, by means of which the enemy may be put to flight and pursued: and of others, more solid, which will resist both fire and sword and which are easily lowered or raised. I know also of a means to burn and destroy hostile bridges.
2. In case of the investment of a place, I know how to drain moats and construct scaling ladders and other such apparatus.
3. Item: If, by reason of its elevation or strength, it is not possible to bombard a hostile position, I have a means of destruction by mining provided the foundations of the fortress are not of rock.
4. I know also how to make light cannon easy of transport, capable of ejecting inflammable matter, the smoke from which would cause terror, destruction and confusion among the enemy.
5. Item: By means of narrow and tortuous subterranean tunnels, dug without noise, I am able to create a passage to inaccessible places, even under rivers.

6. Item: I know how to construct secure and covered wagons for the transport of guns into the enemy's lines, and behind which the infantry can follow without danger.

7. I can make cannons, mortars, and engines of fire, etc., of form both useful and beautiful and different from those at present in use.

8. Or, if the use of cannons happens to be impracticable, I can replace them by catapults and other admirable projecting weapons at present unknown; in short, where such is the case, I am able to devise endless means of attack.

9. And if the combat should be at sea, I have numerous most powerful engines both for attack and defence; and ships that are both gun-proof and fire-proof.

10. In times of peace, I believe that I can compete with anyone in architecture and in the construction of both public and private monuments and in the building of canals.

Milan at this time faced threats from Venice to the east, the Pope's armies to the south and the King of France to the north. Leonardo wrote his letter in the full knowledge that engineering skills and military expertise trumped artistic abilities, despite rich patronage of the arts by wealthy ruling families such as the Medicis in Florence and the Sforzas in Milan. Only at the end of the letter and unnumbered does he add: 'I am able to execute statues in marble, bronze and clay: in painting I can do as well as anyone.' He also offered to create a bronze horse in memory of Sforza's illustrious father. The letter served its purpose: he got the job.

Born in 1452 in the small Tuscan village of Vinci, near Florence, as a result of a relationship between a young lawyer, Ser Piero da Vinci, and a peasant girl, Caterina, he probably enjoyed a happy childhood in the country. In 1469 Ser Piero moved to Florence, probably attracted there by the wealth of the city, deriving from banking and textiles, and enrolled Leonardo as an apprentice to Verrocchio, an accomplished sculptor and artist with strong links to the ruling Medici family. Here he would have been intellectually challenged by the lively minds of the other apprentices, and their experimentations may well have extended to personal relationships; in 1476 four of them, together with Leonardo, were put on trial for sodomy, but acquitted for lack of evidence. More importantly, he profited from close contacts with outstanding minds of the time such as the mathematician Bennedetto d'Abbaco, the Greek scholar and translator Agiropulo, who had fled Constantinople after its fall to the Ottomans, and Paolo Toscanelli, a physicist and geographer.

In Milan he endured, initially, fairly straitened circumstances until granted quarters and a workshop in the castle, allowing him to take on students, notably Salai, who became his long-time companion. He also developed during this time a close friendship with Pacioli, a Dominican Friar and eminent mathematician. When the French entered Milan in 1499, he returned to Florence via Mantua and Venice, taking Pacioli and Salai with him. In 1502/3 he travelled briefly with Cesare Borgia as an adviser on his military campaigns and during this time he met Machiavelli, both of them returning after a short time to Florence. In the ten years from 1503 he spent various periods in Florence and Milan, invited to visit the latter by Louis XII, now in control of the city; and while in Milan in 1513 he received an invitation to come to Rome by the recently installed Medici

Pope, Leo X. He responded positively, taking Salai and other students, notably Melzi, with him. Ignored in Rome by leading figures such as Bramante and the much younger Michaelangelo and Raphael, and given little to do by the Pope's court, he accepted an invitation by the young Francis I, then campaigning in Italy, to accompany him back to France. He spent the next two and a half years in peaceful retirement at Blois, near the royal residence at Amboise, with an ample pension and with occasional visits from the King himself.

Leonardo's fame rests on no more than a dozen paintings and some 7,000 pages of sketches and notes. The paintings need no further comment here, other than to mention the unfortunate deterioration of *The Last Supper*, which commenced soon after its completion, owing to his unsuccessfully experimenting with the pigments and materials for fixing the paint to the walls. Later, exposure to the elements unfortunately caused further damage. The sketches and notes which survive cover a vast range of subject matters, natural and man-made and man-imagined: geology, climate, biology, human proportions and anatomy, architecture and town planning, hydraulics and hydraulic works, and innumerable devices and machines. Many of the devices and machines already existed, either in actual physical form or in sketches by earlier or contemporary engineers such as Taccola and Francesco di Giorgio, but in many cases Leonardo sketched possible variations, extensions or improvements to these. Other innovations in his sketches included the exploded diagram showing the individual elements of a mechanism separately, and both plans and elevations.

His sketches and notes (in mirror writing), committed to paper using red chalk, pen and ink, pencil or charcoal, had no influence on subsequent advances in science and technology because they disappeared from general view after his death. He left them (some 7,000 pages) to his one-time student Francesco Meltzi, with whom he became particularly friendly in his later years. Meltzi guarded them jealously, neither cataloguing nor publishing them, for fifty years until his death, upon which they passed to his son Orazio, who had no interest in them and allowed various people to take what they wanted – notably his house tutor, Lelio Gavardi, who pilfered thirteen manuscripts and tried unsuccessfully to sell them to the Duke of Florence.

Eventually, Pompeo Leoni, sculptor to the King of Spain, acquired a good many of them and had them catalogued, reorganised and rebound, although not in chronological sequence. Many of these were bought from Leoni's heir by an Italian, Count Arconati, who donated a large number to the Ambrosiana library in Milan. Others found their way by sale or gifts to other museums and private persons. Napoleon removed the Ambrosiana collection to Paris, most, but not all, of which were returned to Milan after his death.

Michelangelo Buanarroti (1475–1564)

Sculptor, painter, architect/engineer, poet and excelling in all these: truly a Renaissance Man, with strong claims to be *the* Renaissance Man. Born in 1475 at Caprese, in the Arno Valley, into a highly respected Florentine family, Michelangelo showed great

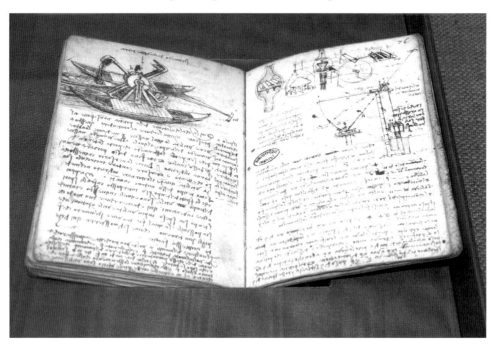

A Leonardo da Vinci notebook.

This sketch of a hoist by Leonardo shows both the assembled contrivance on the left and, innovatively, an exploded diagram on the right depicting its individual elements. Operated by a to and fro motion of the lever, each motion raised the suspended load.

drawing skills from an early age, prompting his magistrate father, somewhat against his inclinations, to apprentice his second of five sons at the age of thirteen to the artist Domenico Ghirlandaio. If one of his skills transcended the others, it was in sculpting and these skills emerged early in his apprenticeship, to the extent that Lorenzo d'Medici, particularly seeking someone with these skills, invited him to join his school at the age of about fourteen to study under Bertoldo di Giovanni. The school folded with the death of Lorenzo in 1492, but not before Michelangelo had carved two important reliefs, a *Madonna of the Steps* and a *Battle of the Centaurs*. He continued to work in Florence, producing notably a carved wooden crucifix for the church of Santo Spirito, until the imminent threat to the city from the French King Charles VIII saw him depart for Venice and Bologna, producing in the latter three carved marble figures for the shrine of St Dominic. With the overthrow of the Medicis and the rise of Savanorola, he had little incentive to make an early return to Florence and the year 1496 found him in Rome, carving for a private client a statue of Bacchus, followed by one of his great masterpieces, the *Pietà*, commissioned by the Pope for St Peter's Basilica. In 1501 he responded to an invitation by the new republican government in Florence to create a statue of David to stand in the Piazza della Signoria, soon to become his most famous and visible sculpture, and commenced work on a commission to paint the Battle of Cascina for the Council Chamber of the Palazzo. He abandoned this commission at the cartoon stage in 1505 to respond to a summons from the recently installed Pope Julius II to create a grandiose sepulchral monument in Rome. Julius immediately dispatched him to the marble quarries at Carrara to supervise the extraction of the large stone blocks and transport them to Rome: an engineering task he accomplished successfully.

But on arriving back in Rome from Carrara, he found that Julius had transferred his interest from his tomb to the building of a new St Peter's for which, to Michalangelo's chagrin, his much older and more experienced rival Bramante had been appointed engineer/architect. In fact, he continued to work on the tomb on and off over many years and with many assistants, completing it in 1545. In 1508 Julius, not wishing to lose him to Florence, offered him the task of painting scenes from Genesis on the vaulted ceiling of the Sistine Chapel, a task he completed by 1512, thereby creating almost single-handedly one of the great frescos in Western art, having entrusted a minimum of the work to his assistants.

By 1516, with the Medici back in power in Florence, Michelangelo returned to that city. Evidently forgiven for having sculpted the statue of David for the republican government, he was commissioned to build the Medici funerary chapel, challenging both his architectural skills for the chapel itself and sculpting skills for the four tombs to be placed in it. He worked on it intermittently until returning to Rome in 1533, with a break from 1527 to 1530, when the Medici were temporarily expelled from Florence. He never returned to Florence, and by the time of his departure from the city had completed only two of the tomb sculptures, with a third partly completed.

In 1529, with Florence under imminent threat of attack from the combined forces of the Empire and the Papacy, the Florentine Council of Ten of War entrusted him '...to be governor and procurator-general over the construction and fortification of the city walls, as well as every other sort of defensive operation and munition for the town of

Florence, paying him one gold florin a day for his services. He fortified the hill of San Miniato and when it came under attack from heavy artillery, he covered the masonry with sacks of wool and mattresses to protect it from the shock of cannon balls. Leonardo had suggested a similar practice in his notes, but it is unlikely that Michelangelo had seen these. Unfortunately, the competence of the military high command in Florence did not match Michelangelo's skills and Florence fell to the allied forces despite the fortification. Taking a pragmatic view, Michelangelo escaped through enemy lines under cover of darkness, taking a large sum of money, presumably a stash of gold florins, with him and headed for Venice. Proposals for replacing the timber Rialto Bridge in Venice had been under discussion since a destructive fire had threatened it in 1512 and Michaelangelo prepared plans for a new masonry bridge to replace it. So did several others. It was not replaced until late in the sixteenth century.

Back in Rome once again, he found his engineering talents in demand. Commissioned to repair the foundations of the bridge of Santa Maria, threatened by scour, he constructed coffer dams to isolate the foundations and had almost finished repairing them when the authorities transferred control of the work to an architect named Nanni di Baccio Bigio. The bridge collapsed in 1557. Pope Paulus IV, embroiled in a war with Spain, placed him in full charge of reconstructing the defences of Rome, but when the enemy advanced on the city in 1557 he once again secretly departed the city, and headed this time for Spoletto. The final design of the dome of St Peter's owed much to Michelangelo, whose actual engineering achievements considerably exceeded those of Leonardo. Although the artist and sculptor transcended the engineer in Michelangelo, he had nevertheless acquired excellent engineering credentials and experience to bring to the problem of building the cupola of St Peter's by the time of his appointment as architect/engineer to the works in 1547 at the age of 72. He died in 1564 at the age of 89. His irascible, acerbic nature throughout his life did not endear him to his contemporaries, but they certainly recognised and exploited his remarkable talents.

Sinan (1489–1588)

Although most noted for his great mosques with their magnificent domes roofing the central prayer areas, the many hundreds of diverse works completed under Sinan's responsibility during his long lifetime included mosques, hospitals, palaces, schools, public baths, mausoleums, bridges, aqueducts, water supplies and sewers. His long life of 99 years spanned the reigns of five sultans, Beyazet II (1481–1512), Selim I (1512–20), Süleyman the Magnificent (1520–66), Selim II (1566–74) and Murat III (1574–95). He built two of his three greatest mosques, the Seyzade and Süleymaniye in Istanbul, in the time of Süleyman; and the third, the Selimiye in Edirne, for Selim II, he completed in the 80th year of his life.

Born in Anatolia into a Christian family in 1489 with the given name of Joseph, he was conscripted by his Turkish captors into the Janissary Corps in Istanbul at the age of 23, and after conversion and circumcision was given his Islamic name of Sinan. Although set up from captured Christian youths as an elite fighting unit, totally loyal to the Sultan,

Janisseries could, in some cases at least, practice a trade as well as undertaking their cadet training. Several years older than the average age at which most youths were conscripted, Sinan had already worked in his village for many years as a stonemason and carpenter and he continued to pursue these after induction into the Corps. His skills and obvious intelligence came to the attention of leading engineer/architects and under their tuition he became an engineer/architect himself. He continued his military career as well, leading to his admission to the brotherhood of Janisseries, and participated over many years in campaigns as widely dispersed as Rhodes, Baghdad, Belgrade, Vienna and Cairo, in the last case, as Chief Architect, being given the authority to demolish any building out of sympathy with the city plan. During his campaigns he constructed bridges, siege engines, cranes, pontoons and ships and built – and advised on the demolition of – defensive walls, towers and other structures. In the Battle of Mohács in 1526, cavalry subaltern Sinan distinguished himself to the extent of being mentioned in the order of the day, which brought him to the attention of the new sultan commanding the troops, Süleyman the Magnificent, and consequent promotion on the battlefield to captain of a company made up of Janissery cadets.

In 1539 the new Grand Vizier, Celebi Lüfti Pasha, appointed Sinan, now 50 years of age, to the exalted position of Royal Chief Architect and Engineer, with the glorious title Architect of the Abode of Felicity (Istanbul), heading a department responsible for all public works in Istanbul, including roads, bridges, water supplies and sewerage, accredited with altogether some 360 works throughout the extensive Ottoman empire. He clearly could not have designed and supervised more than a small number of these himself, but it is a testament to his skill in selecting and training his assistant staff and delegating authority.

In 1583, and now 94 years of age, he made his second trip to Mecca, by ship and camel, this time as a pilgrim, his first having been for the purpose of restoring the arcaded enclosure surrounding the Kaaba stone. He also on this first trip restored damaged aqueducts supplying water to the holy city and added a new aqueduct. He died on 17 July 1588, having had to spend the last few months of his life in a wheelchair, possibly as a result of a minor accident. His tomb, of his own design, stands close to that of Süleyman, just outside the walls of the Süleymaniye Mosque. Its magnificent headstone, carved in the form of a voluminous turban of state, has inscribed upon it a eulogy by the Mustafa Sa'i. Of all the works attributed to Sinan, the eulogy chooses to make specific mention of the bridge at Büyükchekmejé, which Süleyman had ordered to be built after he had nearly drowned in 1563 in the Ringed Valley just inland from Büyükchekmejé. Caught out by a flash flood while hunting for migratory water birds, he managed to shelter in the rafters of his hunting box from the rampaging waters, which destroyed the local bridge. On his return to Istanbul, Süleyman ordered Sinan to build a replacement bridge able to survive such storm waters, and allocated a huge sum of money for the work. In order to solve the problem, Sinan had a number of artificial islands constructed across the estuary and linked these with four humped-back bridges, which to this day are a visual testimony to his skills as an engineer. Interestingly, Sinan chose to sign his name on a post at the end of the bridge as Yusuf (Joseph) Abdullah.

Cornelius Vermuyden (1595–1677)

In 1618 Joachim Liens came to England as one of three commissioners of the Netherlands Government to discuss with the English Government matters relating to the Dutch East India Company and to herring fisheries, and a year later these three worthies became recipients of Knighthoods from King James. Liens came from a prominent family in the Zeeland town of St Maartensdijk (St Martin's Dyke), which had for generations been involved in Dutch polder reclamation, as had the Vermuyden family, also prominent in the town. Close historical links between the families had been strengthened by the marriage of Joachim to Cornelia, the sister of Cornelius Vermuyden, although tragically she had died very young in 1612. His visit to London proved somewhat fortuitous as King James confessed throughout his reign an interest in land reclamation. Most interest by potential investors, both English and Dutch, was focused on reclamation of the East Anglian Great Level, and Joachim kept his brother Cornelius Liens, a possible investor, fully informed, particularly in relation to the King's intentions, at least as far as he or anyone else knew these. Cornelius Liens had, in fact, as early as 1506 presented a proposal to the King for draining the Great Level, but it ran into opposition from Lord Chief Justice Popham, who wanted to undertake the project himself.

Cornelius Vermuyden's exact birth year in St Maartensdijk is uncertain, but he is believed to have been about 26 years old when he left Holland in 1621, perhaps at the instigation of Joachim Liens and others in the Netherlands looking to invest in land reclamation schemes in England. Little is known about his activities up to this time, although he seems to have been employed as a tax collector for a time. Living in St Maartensdijk, he would certainly have been familiar with land reclamation from birth, as it would have been a recurrent household topic at the time, and the fact of his having been invited to England, whether by potential investors or possibly by the King himself, suggests he had in fact made something of a name for himself in this field of endeavour. His decision to accept the invitation may have been strengthened by the end of the twelve-year truce between Spain and the Netherlands, potentially limiting his likely opportunities in his homeland.

His first enterprises in England commissioned by the King proved to be of fairly limited magnitude, and included a small job on a breach in the Thames bank at Dagenham and some drainage work in Windsor Great Park. His big break came with the death of James I in 1625 and the accession of his son, Charles I, who signed an agreement in 1626 for Vermuyden to drain the 70,000 acre Royal Chase at Hatfield on the south Yorkshire–Lincolnshire border, to be financed by Netherlands investors in return for ownership of one-third of the reclaimed land, including an allocation of 4,554 acres to Vermuyden as principal undertaker. His employment of Dutch and Flemish workers with experience in land drainage as his main workforce caused an outcry in the local population, despite one-third of the reclaimed land being allocated for common rights of pasture. As a result of damages wrought by local people, requiring expensive remedial excavations and embanking, and various lawsuits, the investors lost most of their money on the enterprise. Technically, the works were a partial success and for Vermuyden himself the lessons he learnt proved valuable in undertaking the much larger project of draining the Great Level of East Anglia. Unlike his investors, he profited too, enabling him to buy from the Crown Malvern Chase in Worcestershire for £5,000, which he reclaimed and enclosed, to take a grant of 4,000 acres of land at Sedgemoor in Somerset for £12,000 and with Sir Robert Heath, Attorney General to Charles I, to take a lease for 30 years on the Dovegang lead mine in Wirksworth in Derbyshire. The King bestowed a knighthood on him in 1629 and he became a British citizen in 1633.

Vermuyden gained little financially from his work on the Great Level, carried out in two stages: pre-Civil War, 1630–7, and post-Civil War, 1649–53, due to underfunding, litigations and problems with local peoples and vested interests. At the end of the work he seems to have been left with little more than his Sedgmoor land and his interest in the Wirksworth lead mine. He retained these throughout the remainder of his years, about which little is known, but he appears to have led a quiet retired life, taking some satisfaction from the attainments of his children. He had thirteen children between about 1624 and 1644, at least two of whom died in childhood. His oldest son Cornelius became a founder member of the Royal Society in 1663. One of his daughters married a Baronet and three others married into landed gentry. The youngest son, Charles, graduated from Christ Church, Oxford and became a licentiate of the College of Physicians in 1662. He married into a landed Devonshire family, but died young in 1673.

Pierre Paul Riquet (1604–80)

In 1662 Colbert, Louis XIV's Controller General of Finances, received a letter from Pierre-Paul Riquet, a wealthy landowner in the south of France. Although lacking any engineering training, and already middle-aged, he sought to take on the construction of a canal to link the two great oceans, the Atlantic and the Mediterranean, which washed the shores of France. His spur at this time may well have been the realisation that the completed canal would enhance the value of his property, but as events later showed his obsession with the project overrode this consideration. His professional life as a tax-gatherer had clearly honed his diplomatic skills as well: he enlisted the support of Charles d'Anglure de Bourlemont, the aristocratic and influential Archbishop of Toulouse, as evidenced in his letter:

> Monseigneur
> I am writing from this village about a canal that could be constructed here in the Province of Languedoc, joining the two seas. You will be surprised to hear me speak of something of which apparently I know nothing – one does not expect tax gatherers to go about with levelling instruments. But you will excuse my daring when I explain that I am writing at the orders of Monseigneur the Archbishop of Toulouse. Some time ago the Archbishop did me the honour of visiting me, and he asked how the canal could be constructed, having heard I had made a particular study of the problem …

He continued at some length, extolling the advantages of such a canal and stressing its benefits, including those likely to appeal particularly to Colbert and the King, including the advantages of connecting the Atlantic and Mediterranean, by-passing Gibraltar and thus reducing the revenues of the King of Spain. The Archbishop of Toulouse also wrote to Colbert expressing his strong support for the scheme.

The possibility of building such a canal had been lodged in his mind from his youth, it having been a topic of discussion in his parents' household; although his father, an influential and wealthy lawyer, had in fact voted with the majority against an early proposed scheme being considered by his local council, probably because it passed through Narbonne rather than his home town of Bèzier. Born in Bèzier in 1604, to a family with distant Italian Gibelin origins, Pierre Paul demonstrated his loyalties by preferring to speak in la Langue d'Oc, from which the province derived its name. Educated at the Jesuit College in Bèzier, his talents lay not in the study of languages, but in the sciences and mathematics, the latter pleasing his father who saw him as a future financier and man of business. At nineteen years of age he married Catherine de Milhau, daughter of a local wealthy family, whose dowry enabled him to buy the chateau and estate of Bonrepos, situated on the slopes of a valley twelve miles to the east of Toulouse. Later in life, when ennobled by the King he chose the title Baron Riquet de Bonrepos.

A strong influence in his early life, his godfather Father Portugniares in 1630 helped secure for him the lucrative position of collector for Langedoc of the salt tax, which had become one of the state's chief sources of revenue since its first imposition in 1206. Within a year or two his responsibilities for collecting this tax extended to the whole province and to help him he appointed in 1632, as his deputy, an ex-school friend and future brother-in-law, Paul Mas, a Doctor of Law practising his profession in Bèzier. As a result of his extensive tax collecting travels within the province, he acquired an intimate knowledge of the countryside, which held him in good stead in presenting to Colbert and the King the proposal for a major canal passing through it. With the death of his father in 1632, Riquet inherited much of the family estate, which enabled him to make the house in Bèzier the hub of his business activities, which included a very profitable military contracting operation supplying the King's armies in Cerdagne and Roussillon.

His commitment, to the point of obsession, with the canal no doubt helped persuade a royal commission appointed by the King to examine its merits that it should be built, but its construction was to have disastrous consequences for his finances and his health. Finances from the bridge came in part from the Royal Treasury (sometimes only after a protracted interval) and in part from Riquet himself, on the agreement that certain local taxes should be assigned to him for a period of six years. These turned out to be less than expected. He died in 1680 at the age of 76, a few months before the canal opened, deeply in debt and after several years of ill-health, brought on by worries about slow progress, overcoming physical obstacles, dealing with difficult landowners, finding sufficient bricks for his locks and quarrels with the commissioners about the route.

The ownership of the canal passed to his two sons in the ratio of two-thirds to the elder Jean Mathias, who had been actively involved in the canal construction with his father, and one third to the younger Pierre Paul II, a French army officer who had had no involvement. They first had to discharge the huge debt incurred by their father and to do this sold over half the shares in the canal, but with the option of being able to repurchase them when able to do so. The success of the canal did enable the family to eventually repurchase the shares and they became very wealthy as a result.

Christopher Wren (1632–1723)

In 1669 the dean and archbishop, with the approval of the King and the rest of the Commissioners, appointed Christopher Wren to the post of Surveyor to be responsible for the rebuilding of St Paul's Cathedral on the site of Old St Paul's, largely destroyed in the Great Fire of 1666. The term Surveyor encompassed in one person the role of Architect, Structural Engineer, Foundation Engineer and, as necessary, various other specialisations which would be employed today in such a major undertaking. This breadth of responsibilities would have presented little problem to a man whose interests, and in some cases expertise, extended over a wide range of disciplines in addition to building construction and design, these including astronomy, anatomy, optics, mechanics, meteorology and even road paving. His skill with instruments and particularly model building held him in good stead in presenting his plans for the new St Paul's. He also had the option of consulting with his friend Robert Hooke if need be.

Christopher Wren was born in Wiltshire in 1632, into a family with strong connections to the Church; his father, also Christopher, in 1635 succeeded his own brother Matthew as Dean of Windsor. The family had strong Royalist leanings and Matthew's devotion to the King earned him successively the senior Church positions of Bishop of Hereford, Norwich and finally Ely. It also saw him imprisoned in the Tower of London for 18 years from 1642 to 1660. Christopher senior, meanwhile, chose to adopt a lower profile and moved from Windsor with son Christopher to live quietly at the rectory in Bletchingham, Oxfordshire, with his daughter Susan and her husband and rector William Holder, a mathematician and graduate of Pembroke College, Cambridge. William was to have a strong influence on the young boy, tutoring him in mathematics and interesting him in scientific matters and astronomy.

In the early 1640s Christopher entered Westminster School in London under headmaster Dr Richard Busby, a strict disciplinarian, who favoured for admission children of families known to have strong Royalist leanings. Here he received an excellent grounding in mathematics and science as well as the Classics; but in 1646 his father withdrew him from the school, perhaps because of his now straitened circumstances and perhaps concerned too with his safety in a school where Busby maintained his strong Royalist bias despite Parliamentary success in the Civil War. Wren remained in London for most of the next three years, in the household of physician Dr Charles Scarborough as his assistant on anatomical experiments, but able, too, to pursue other scientific interests, including experiments with sundials. He entered Wadham College, Oxford, in 1649, receiving his BA degree in 1651 and MA in 1653, becoming a fellow of All Souls College in the same year at the early age of 21. Temporarily deserting Oxford, he accepted in 1657 the post of Professor of Astronomy at Gresham College in London, but returned to Oxford in 1661 to take up the post there of Savilian Professor of Astronomy. Throughout these years his interests encompassed, as well as astronomy, various mathematical and geometrical problems, and such diverse matters as ways to find longitude at sea, surveying instruments, machines to lift water, military devices for defending cities and means for fortifying ports.

Regular weekly meetings he initiated with like-minded men to discuss these emerging scientific subjects led eventually to the establishment in 1662 of the Royal Society, England's premier scientific body, of which he became President for the period 1680–2. He not only included architecture in his many areas of interest, but obviously harboured for some time a wish to make his own contribution in this area. His uncle Matthew Wren, Bishop of Ely, knew of this or was certainly made aware of it and may even have initiated the idea in his nephew's mind. Bishop Wren had vowed to devote a sum of money to some holy and pious purpose on his release from incarceration in the Tower of London, a vow that he made good in the form of a new chapel for Pembroke College in Cambridge, where he had been an undergraduate and elected a Fellow in 1605. He employed his nephew as architect for the new chapel, completed in 1665, Christopher's first architectural commission, resulting in a religious building unique for its time in the absence of Gothic features. His next commission was the grander Sheldonian Theatre in Oxford and others soon followed. He missed the consecration ceremony for the Pembroke chapel in September 1665, presided over by his 80-year-old uncle, having crossed the channel to study various building projects being carried out in Paris by Louis XIV and his Minister of State, Colbert. He stayed there for about six months and also took the opportunity to meet with leading French scientists and the famous Italian sculptor and architect Bernini, himself on a brief visit to the city at the time.

In 1663, three years before the Great Fire of London, Charles II had set up a Royal Commission to examine St Paul's Cathedral, the largely medieval structure of which was by then in a dilapidated state, and make recommendations on repairs and restoration work. The commissioners sought the advice of a number of architects, including Wren, despite his having no completed building to his name at this time. A committee appointed by the commissioners met on site one week before the Great Fire to assess the problems of the old cathedral and to hear from three architects, including Wren, their proposals for its refurbishment. Impressed by Wren's enthusiasm and his clear understanding of the problem, and seduced by his proposal to replace the tower with a dome, they decided to recommend Wren's scheme for the structure to the commissioners. But a week later everything changed.

Less than a week after the fire, Wren presented to the King a scheme not just for rebuilding the Cathedral, but for the whole burnt-out area of the city, based on a great oval piazza into which nearly all principal streets converged. Other schemes soon followed and despite the King favouring Wren's proposals, various opinions stemming from Parliament and a Rebuilding Committee set up by Charles II delayed a decision, further delayed by war with the Dutch, whose ships patrolling the mouth of the Thames had stifled trade with the city.

Wren's appointment as surveyor for St Paul's came in 1669, and in the same year the monarch made him Surveyor General of the King's Works, in charge of the reconstruction of London with the invaluable assistance as Surveyor, with two others, of Robert Hooke, a fellow member of the Royal Society and an outstanding experimentalist, who had presented plans of his own for reconstructing the city. Opposition by vested interests, particularly property owners, scuppered any possibility of Wren's Grand Scheme being realised and the medieval street layout prevailed. This had the benefit that the new Cathedral would remain the focal point of the city, which would not have been the case if his Grand Scheme had been implemented. Wren's reconstruction included fifty-one churches and other important public buildings, and other public works of an engineering nature such as paving streets, and relaying sewers, water pipes and conduits. The work also included the excavation of the malodorous Fleet ditch to create a canal navigable at high tide from the Thames to Holborn bridge, with retaining walls, quays, warehouses and bridges. In an echo of, and possibly inspired by, Alberti, Wren introduced inverted stone arches into the foundations of Trinity College Library in Cambridge to combat the danger of differential movements deriving from marrying new works with old remnant foundations.

Although 43 years old at the time of commencement of construction of St Paul's, Wren lived to see its completion 35 years later and did not die until 12 years later at the advanced age of 90 years.

Sèbastien Le Prestre de Vauban (1633–1707)

In the seventeenth century the emphasis in fortress construction switched from Italy to France, which country produced perhaps the most brilliant military engineer of all time, his genius extending to both attack and defence. Born in the small village of Morvan in 1633, the son of an apparently impecunious country squire in Burgundy, Sébastian Le Prestre Vauban spent much of his boyhood playing in the woods with the local peasant children. At the age of ten he entered the Carmelite College at the small town of Semur-en-Auxois to be tutored by the Abbè de Fontaine, and where he showed particular aptitude for mathematics, science, history and drawing. He may also have spent time studying the walled ramparts of the town. He began his military career at the age of seventeen, accepted as a Cadet in the Regiment of Louis II of Bourbon, Prince de Condè, and thus became involved in the battles of the Parlement and Paris mobs, known as the Fronde, against the Chief Minister Mazarin, who had the support of Anne, widow of Louis XII and Regent for her young son Louis XIV. Draconian taxation measures introduced by Mazarin to pay for the war against the Habsburgs, and a general dislike for the man, born in Italy, triggered the uprising. Mazarin had Condè imprisoned in 1650, fearing that he might support the Fronde; but Anne, mindful of Condè's popularity, had him released, whereupon he embarked on a number of sieges against the supporters of the Crown. Mazarin, meanwhile, although having fled the country, came to an agreement with Turenne, a leading royal general, who saw off Condè by 1653 and helped re-establish royal absolutism. Louis XIV, now of age, re-established Mazarin as his Chief Minister.

Vauban distinguished himself on a number of occasions fighting with Condè's forces, and when taken prisoner by royalist troops was quickly released; Mazarin, recognising his outstanding qualities, offered him a commission. He became an Ingènieur Ordinaire du Roi by the age of 22 and proceeded to serve his country with distinction in its war against Spain. His military expertise extended equally to siege methods and to innovative defensive fortress design. In 1672 he proposed that the engineers in the army should become an organised special branch of the services and this resulted in the establishment of the Corps Impèrial du Gènie. In 1678 he became Inspector-General of the French Fortresses and in 1703, four years before his death, Maréchal de France. After fifty-five years of military service, he had his last command at Dunkirk in 1706.

Although strictly a military man, his responsibilities, in common with other French military officers, extended to civilian projects, particularly where these may also have had some military significance. His work on the construction of inland waterways and the canalisation of rivers, while serving the commercial interests of the country, also provided a flexible and rapid movement of equipment in times of war. He had a minor involvement in the construction of the Langedoc Canal, including in 1686/7 the excavation of a 7 metre channel and 120 metre long tunnel to augment the flow of the Laudot River, which supplied water to the St Ferreol dam.

A matter of great concern to Vauban was the suffering of millions of his countrymen under the tax system of the King's administration, the gross unfairness of which he witnessed time and again during the extensive and endless travels through the Kingdom made necessary by the carrying out of his duties. Year after year, the administration's expenses exceeded receipts; in 1683, for example, expenditure amounted to 109 million livres with revenues only 93 million livres, while in 1697 after the War of the League of Augsberg the relative figures were a staggering 219 million and 81 million. The Finance Ministry's tax collectors resorted to unscrupulous methods, with the burden falling on those least able to pay; one of the most pernicious measures was a tax on salt, a certain quantity of which everyone was required by law to purchase. In 1706 Vauban published his scheme entitled the *Dîme Royal*, proposing a much fairer system of taxation. This publication earned the wrath of many vested interests and possibly that of the King himself. Suffering from pneumonia, he died a few months later, ignored and unnoticed, his enormous contributions to the welfare and safety of France deliberately ignored and counting for nothing. A hundred years later Napoleon, fully cognisant of Vauban's place in the history of his country, had his heart re-interred in Les Invalides in Paris in a formal ceremony attended by Vauban's descendants, senior Ministers and other important persons.

Jean Rodolphe Perronet (1708–94)

The word 'perron' means a large stone and Perronet's name could hardly have been more appropriate for a man who spent much of his life building masonry bridges, and in the process brought their design to the point of perfection. And how ironic, then, that the eighteenth century virtually saw the end of masonry bridge construction, as first iron, then steel, became the dominant materials for major bridge construction in the nineteenth century, with reinforced concrete to follow. He not only built bridges but wrote about them, and his *Traité de la Construction des Ponts* is regarded as a classic.

Born in the small town of Suresnes, near Paris, in 1708, the son of a Swiss soldier in the service of France, he had the unusual experience at six years of age of meeting the four-year old boy shortly to become the next King of France, Louis XV. While playing in the Tuileries, he caught the attention of the young prince, playing in adjacent gardens, no doubt heavily protected and eager for someone to talk to and to share in his games; in the event Louis managed to persuade his minders to invite Jean Rudolphe to join him and this proved to be the beginning of a close friendship between them. Louis XV, at five years of age, succeeded his great-grandfather Louis XIV, who died in 1715 after ruling for seventy-two years, but was not crowned until 1723, his great uncle Philippe II, Duke of Orlèans, serving as Regent until Louis came of age.

At 17 years of age, Perronet joined the staff of an architectural firm in Paris specialising in public works, where he remained until, at the age of 37 in 1745, he entered the Corps des Ingénieurs des Ponts et Chaussées and a year later became engineer-in-chief for the Alençon area. This short-lived appointment lasted only a year when, in 1747, he was appointed Director of the newly established Ecole des Ponts et Chaussées, set up to train civil engineers to serve in the Corps. In 1750 he assumed the role of Chief Inspector, and in 1763 Chief Engineer of the Corps. His first major bridge assignment came in 1763 on the Seine at Mantes, where the first river pier, slim by contemporary standards, on placing the keystones of the first arch showed signs of tilting, which he corrected by placing the voussoirs of the second arch 'avec la plus grand célérité'. He went on to design a number of great masonry bridges with flat arches, with self-equilibrating thrusts carrying through to the abutments, and having slender piers giving least resistance to river flow. Interestingly, the three flat segmental arches of the Ponte Vecchio in Florence built by Taddeo Gaddi three centuries before would have behaved in the same way, at least in some measure, although the span to pier width ratio of five, remarkable for its time, was much less than that of ten achieved by Perronet.

Notwithstanding the success of his existing structures, Perronet met resistance in his design for his last great bridge, the Pont de la Concorde. Attacked from a number of quarters, he resisted a call to increase the pier widths, but made them solid rather than his chosen design of having them divided and joined by a lateral arch, and increased the rise of the spans by up to a metre. He also provided an open, balustraded parapet to increase the sense of lightness. It was during the construction of this bridge, started in August 1788, that the Revolution broke out, but fortunately did not hinder its progress, the elderly Perronet having installed himself in a hut at one end to personally supervise the work through to its end in 1791.

Perronet's interests embraced hydraulic engineering and related machinery, such as devising water wheels worked by the river itself to operate pile drivers and to pump water out of cofferdams. His enormous contribution to the profession of the civil engineer cannot be overstated, through his enlightened Directorship of the Ecole des Ponts et Chaussées which, together with the Ecole Polytechnique, has ever since furnished France with outstanding civil engineers over the years. One of Perronet's own students, Emiland Marie Gauthey (1732–1806), rose to become Inspector-Générale des Ponts et Chaussées in Paris and taught for a time at the Ecole des Ponts et Chaussées. A builder of numerous bridges, he also produced the Traité de la Construction des Ponts, although this was not published until after his death.

Perronet died in 1794, three years after the completion of the Pont de la Concorde.

John Smeaton (1724–1792)

In 1753, when John Smeaton undertook his first construction project, the Houghton Mill in Lancashire, water was the only available major source of power other than muscle and, to a lesser extent, wind power. The design of watermills remained his first love throughout his life: he designed at least fifty in his lifetime and his 1759 paper to the Royal Society, *An Experimental Enquiry concerning the Natural Powers of Water and Wind to turn Mills, and other Machines, depending on a Circular Motion*, won the prestigious Copley Medal awarded by the Society. In the 1770s he produced his first steam engine designs, a source of power which soon rendered watermills obsolete and more than any other single factor brought about the Industrial Revolution.

Born at Austhorpe Lodge near Leeds in 1724, the son of a well-to-do attorney, William Smeaton, John Smeaton enjoyed a country life as a boy and an excellent education at the Leeds Grammar School from the age of 10 to 16. From an early age he showed a great interest in devices and contrivances and how they worked. By the age of 18 he had become adept in the use of many different tools; he could forge and fuse metals, and had built his own lathe on which he could turn out articles in wood and ivory for his friends. In 1741, at the age of 17, he made the acquaintance of Henry Hindley, a clockmaker with a touch of genius. It was a fortuitous meeting for Smeaton and they became firm friends.

The year 1742 found him in London, encouraged by his parents and paid for by them, to pursue a legal education and they must have been disappointed when, in 1741, he abandoned his law books and returned home to devote his time to building up a set of instruments and making various pieces of apparatus for Benjamin Wilson, who had also abandoned a law career and became a noted experimentalist in electricity. They also shared an interest in astronomy and Smeaton described in correspondence with Wilson how he ground and polished his own lenses to make a telescope. In 1748 he set off again for London to pursue a career as an instrument maker, with his parents' blessing. His breadth of expertise by this time included the design of an improved vacuum pump, precision lathes, telescopes and a mariner's compass, which he tested himself by participating in sea trials with the Royal Navy. He also pursued his classic experiments in the power of water and wind which eventually led to his 1759 paper to the Royal Society, and the Copley Medal. In 1753 he was elected to a Fellowship of the Royal Society.

He obtained his first commission to build a watermill, at Halton in Lancashire, despite having no experience as a millwright; it was possibly offered to him in the light of his experiments on wind and water. This may have had some influence on his decision at this time to devote his future to civil engineering works. Within a year he had produced designs and drawings for a land drainage and reclamation scheme at Lochar Moss in Scotland and plans for the projected Blackfriars Bridge, a commission for the latter ultimately being awarded in 1759 to Robert Milne. In order to broaden his knowledge of civil engineering works he undertook, in 1754, a five week trip to Belgium and Holland to study major hydraulic works in those countries. His major break as a civil engineer came with the destruction by fire in December 1755 of the second Eddystone Lighthouse, the first having been destroyed by a storm in 1703. Surprisingly, with his limited experience in civil engineering works and the enormity of the challenge, the shareholders offered him the commission to build the replacement structure. He spent much of his time over the next four years in Plymouth and successfully saw the project through to its completion. During the winter months, however, when weather prevented work on the Eddystone rock, he investigated and reported on other proposed projects, not least a scheme to render navigable an 18-mile stretch of the River Calder in Yorkshire, requiring seventeen locks. Although put in full charge of its construction from 1760 to 1764, he adopted for this and future projects a consulting arrangement of producing a design report and drawings after assessing the project and examining the site, then leaving construction responsibility to a resident engineer, making site visits as necessary. This enabled him, during the four years of the Calder project, to prepare designs for a number of mills and masonry bridges, draw up plans to improve the navigation of the River Witham and the Fossdyke Navigation in Lincolnshire and for land drainage schemes in Yorkshire, prepare preliminary designs for the Forth & Clyde Canal, advise on securing the foundations of London Bridge and report on the likely effect on the operation of the water wheels of proposals to improve and widen the bridge, and publish a pamphlet on his proposed design for Blackfriars Bridge.

He operated his consultancy from his home at Austhorpe. In 1760 he took on as his pupil the 14-year-old William Jessop, who subsequently became his first assistant and ultimately an outstanding civil engineer in his own right. Smeaton's acting as a consultant in this way, together with specifically designating himself as a 'Civil Engineer' on the title page of his report *Review of Several Matters Relative to the Forth and Clyde Navigation* in 1768, contributed largely to civil engineering now becoming recognised as a profession, no less than that of a doctor or a lawyer. In order to strengthen this image, and no doubt to profit from an exchange of experiences and thoughts in a congenial atmosphere, Smeaton, together with some of his fellow engineers, set up in 1771 the Society of Civil Engineers in the typical style of an eighteenth-century dining club. It still exists today as the Smeatonian Society.

Apart from a three-year period, 1784–7, Smeaton remained active for the rest of his life designing bridges, mills, harbours, river and canal navigations, fen drainage schemes and even steam engines. The failure of Hexham Bridge in 1782 (his only bridge in England, as opposed to several in Scotland) as a result of an exceptional flood undermining the foundations, resting on gravel set into the river bed, followed by the death of his wife in early 1784 at the age of 59, seems to have driven from him the will to carry on. Fortunately one of his two daughters, Anne Brook, took over the household at Austhorpe and by 1787, with many urgent requests for his services piling up, he felt able to resume his activities, but doubled his fees to keep his workload to manageable levels. His largest project during the last five years of his life, Ramsgate Harbour near the mouth of the Thames, he undertook, unusually, on a salaried basis, the Trustees paying him £250 a year. He died at 69 years of age on 28 October 1792.

2

Materials and Methods

A vast variety of the planet's natural resources – earth, rock, minerals, plant and animal life – have been used historically by man to create his works and structures, sometimes used in their natural form, sometimes in an altered form, sometimes in combination and sometimes in a form derived from the naturally occurring material. Sometimes removal of the in situ material comprised the main works, such as in the excavation of drainage or irrigation channels, with the possibility of the excavated material being used to form compacted embankments flanking the channel and thus increasing its effective capacity.

All land-based structures are supported in some way by the ground underneath and it should be given as much attention by designers as the structure itself. The ground becomes, in effect, an integral part of the structure. If the structure is to remain stable or able to perform its function properly the inevitable distortions in the ground caused by the imposed loading must be kept within a tolerable level, which can vary greatly according to the nature and purpose of the structure. If foundations are resting on fine-grained clayey or silty soils the deformations can continue over long periods of time, amounting to decades or more, as water is gradually squeezed out of the intergranular pores. This consolidation, or squeezing out of the water, has the positive effect of gradually strengthening the soil. Many of the great structures, notably the cathedrals, built during the Middle Ages benefited from their protracted construction periods, often amounting to several decades, which allowed builders to correct distortions within the structure, arising from settlements, as construction proceeded and as the soil gradually strengthened below the foundations.

Earth and Rock

Earth exists in many forms, such as clay, silt, sand, gravel, boulders and decomposed chalk or other rock rubble, and since ancient times has provided man with a readily resourced material for the construction of barrows, ritual mounds, embankments, causeways, defensive mounds, walls and ramparts, land reclamation, water retaining dams and road pavements. It has also been used as an aid to construction, for example in the form of temporary ramps for raising large stone blocks in masonry construction. In some instances, enormous amounts of earth have been excavated, transported and placed almost wholly by manpower, with no mechanical assistance, an overt example being Silbury Hill in England, the largest ancient artificial mound in Europe, constructed 4,500 years ago and containing 300,000 cubic metres of mostly chalk rubble, placed by human hand. It is probably waste material originally used for ramps, embankments and causeways in the construction of nearby Avebury, the largest stone circle in Europe.

The pre-industrial age saw the construction in many European countries of structures for the retention, regulation or diversion of water for purposes as diverse as irrigation, fish ponds, domestic supply, canals, driving water wheels, mining and industrial activities. The nature of these structures varied greatly, ranging from embankment dams composed entirely of earth fill materials to structures made up entirely of cut masonry, and in between many of composite design, with cut masonry facings or cores providing the water tight barrier, supported or bulked out with earth or broken rock fill, a type favoured by Moslems in Spain to provide water for irrigation and to drive water wheels. A crude example of a diversion weir 400 metres long crossing the River Garonne in France around 1170 consisted of two rows of lattice work supported by vertical posts, with the space between them filled with earth and stones. Liable to damage, or even demolition, by flood waters, it would have needed frequent maintenance and repair and even to have been rebuilt on occasion. Embankment dams had the advantage that the construction material could be excavated nearby using only pick and shovel and could be taken to site by hand or by simple wheeled handcarts. An English visitor to the site of the St Ferreol Dam in France during its construction wrote that it required to build it an 'incredible quantity of earth, which 2000 woemen daily carry'. Examples of this type of dam constructed to generate water power for mining in the German Hartz Mountains in the sixteenth to eighteenth centuries ranged up to about 15 metres in height, with steep slopes of 1:3 to 1:5. Turves lining the upstream face, or as vertical cores within the structure, provided a degree of watertightness and a wooden, later masonry, spillway at one end took the overflows. Fish ponds, built largely at the instigation of the monasteries and the nobility, provided a continuous and reliable source of food. The dams were invariably of the embankment type, commonly up to 10 metres high, and some may even have conformed to relative dimensions recommended by John Taverner, Surveyor General of the King's Woods south of the River Trent in England, in his published work of 1600, *Certaine Experiments Concerning Fish and Fruite*. He stipulated that the crest width should equal the height and that the base should be three times wider.

The soil or rock underlying the foundations of a structure deform under the imposed weight and it is the responsibility of the designer to ensure that it should not, in itself, suffer catastrophic failure or, short of this, should not deform to an extent that the structure fails or experiences unacceptable distortions of either a visual or functional nature. Many soils, however, behave in an extremely complex manner and furthermore can be influenced by unanticipated events during the lifetime of a structure. The subsequent erection of a nearby structure can have an influence, as can nearby excavations related to such structures or made for other purposes. A rise in the water table can cause severe settlements or collapse in sands, as can earthquake activity. In clay soils, wetting up through a rising water table or the rupture of service pipes can cause swelling or heaving, while drying out through spells of hot weather, exasperated by tree roots, can cause settlements. These heaving or settlement movements occur unevenly and distort the supported structure.

Fine-grained soils such as silts and clays can exhibit long-term time-dependent behaviour. In their natural state, these soils at foundation level are often saturated, or nearly so: that is, the voids or pores between the soil particles are filled with water. When a load is first applied, unless applied very gradually, it is the water in the voids which initially carries the load, or a substantial part of it, through a build up in pore water pressure. Over a period of time, which can be years or decades in some cases, the pore water pressure dissipates (i.e. water is squeezed out of the soil – a slow process, as these soils have low permeability) and the load is gradually transferred from the pore water to the soil structure itself, causing the soil to compress and the incumbent structure to settle, usually very unevenly. In soft clays and silts, these settlements and the resulting distortions to the building can be very large.

Prior to the publications by Karl Terzaghi, the 'Father of Soil Mechanics', and others in the 1930s and subsequently, the engineering behaviour of soils at a fundamental level was not understood, reliable analytical methods did not exist and neither did the knowledge of appropriate tests to determine the relevant properties to put into design and analysis. Experienced builders and engineers would obviously have acquired some appreciation of soil behaviour, but once they stepped outside an area familiar to them they would have been challenged, as soils can vary greatly in their type and characteristics, even within small areas. While failures and untoward happenings must have been commonplace, and often remedied during construction with a minimum of publicity, one factor ameliorated this to some extent – the slow rates of construction in this pre-mechanical age, with reliance mostly on human and other animal muscle power, allowed the underlying soil to gradually strengthen and become better able to support the growing structure. This would have applied both to structures and to embankments for water retention and control, often having to be placed on soft ground.

Building of the great Gothic cathedrals in most cases proceeded slowly, commonly taking thirty years or more, sometimes much more, to complete, thus giving time for any settlements to be absorbed and, if necessary, for remedial measures to be taken in the construction process. Potential problems arose with the subsequent addition of heavy features such as crossing towers or end towers, whether towards or at the end of construction of the main body of the cathedral or, as in some cases, decades or more after its completion. Structural distortion caused by the greater settlements of such towers can be seen in a number of the great cathedrals, and some have required expensive remedial measures such as the underpinning works carried out at Peterborough Cathedral in 1845, at Salisbury from 1859 onwards and York Minster in the 1970s. Winchester Cathedral suffered extensive damage resulting from differential settlements arising from part of the structure being piled, with the adjacent integrated portion resting on spread footings on the soft compressible clays and peats.

While the foundation conditions often seem to have received relatively scant attention by some builders, one writer, at least, bucked this trend. Joseph Moxon, writing in the seventeenth century, recommended 'the Master-Bricklayer, or else his Foreman (which ought to be an ingenious Workman) must in the first place try all the Foundations, in several places, with an iron Croe, and Rammer, or, indeed a borer such as Well-diggers use, to see whether the Foundations are all sound, and fit to bear the weight which is to be set upon them'. If not, he suggests several possible solutions, including ramming in a course of large stones, spreading a foot on each side of the wall, or for looser soils first laying pieces of oak across the trench, about 1 foot apart, and planking over them to the width of the wall. For very loose soils, he recommends driving down piles made of heart of oak, long enough to reach firm ground, with planks spiked to the piles and set as close as possible. For soils only loose in parts he recommends the use of arches spanning over the softer areas.

In his *Ten Books of Architecture* written in the fifteenth century, Alberti stresses the need to know and understand the nature of the soil below the structure in order to provide suitable foundations. As a preliminary assessment, he suggests rolling a heavy weight along the ground or dropping it from a height and observing the behaviour of water in a bowl placed on the ground: if it did not stir, a good foundation was assured. He also recommends digging down exploratory holes to inspect the various soil strata and to identify that which would safely support the structure. He stresses, too, that particular attention should be paid to the presence of water and:

> In marshy Grounds you should make your Trench very wide, and fortify both sides of it with Stakes, Hurdles, Planks, Sea-weeds and Clay, so strongly that no Water may get in; then you must draw off every drop of water that happens to be left within your Frame-work, and dig out the Sand, and clear away the Mud from the Bottom till you have firm dry Ground to set your foot upon.

Timber

Timber was the most widely used structural material throughout the pre-industrial age, in part because of its ready availability, but also because it could be worked relatively easily and offered excellent strength properties in tension, compression and bending. As well as providing primary structural elements, it also found use in secondary purposes such as scaffolding, centring for masonry arch construction, ramps or walkways. Other secondary usage, notably in Byzantine architecture, included its insertion as tie beams in masonry or brick structures to improve resistance to earthquake. The hardwoods birch, oak and beech, and the softwood pine, comprised the timbers most commonly used in construction in northern Europe, oak being the most favoured timber for heavy structural duties, but in some areas where its widespread use led to it becoming scarce, other timbers such as elm replaced it. Elm found particular application in its own right in wet situations and was commonly used for piles: a 1350 inventory of stores kept for the maintenance and replacements for Old London Bridge included 120 pieces of elm with a value of 2 shillings each.

The timber was usually used while still green and without seasoning, mainly because it lent itself more easily to being worked in that condition and perhaps, as an additional advantage, any subsequent shrinkage after installation in a framework or bridge could have had a beneficial effect in tightening up the structure. In general oak logs could be obtained at reasonable cost and effort up to about 9 metres in length, but occasionally greater lengths were sought and obtained at great cost and additional effort for special applications, a notable example being the lantern at Ely Cathedral, built between 1328 and 1342 to replace the great Norman crossing tower that collapsed in 1322 when one of its four supporting piers settled more than the other three, setting up localised stresses that the masonry could not sustain. The cathedral sacrist, Alan of Walsingham, proposed replacing the tower with a remarkable octagonal timber lantern supported on stone piers some 30 metres high, the major structural elements of the lantern itself consisting of eight huge timber posts up to 19 metres high and over 1 metre in diameter. Dispensing with the four support towers of the collapsed square tower, Alan supported his octagonal lantern on the eight end piers of the nave, the choir and the transepts, which he replaced on deeper, more solid footings. Entrusted with the work by the Cathedral authorities, Edward III's master carpenter William Hurley constructed a timber structure at the tops of the piers consisting of straight horizontal members and curved propping members reaching inwards to a horizontal ring, beams forming the base of the lantern. The eight vertical 19 metre high posts of the lantern rise from the nodal points of the timber ring and are given additional support at about mid-height by long inclined timbers springing from the piers. Despite a widespread search, and apparently with no expenses barred, Alan failed to locate eight posts of full length and had to scarf two or three metres to two available posts to reach the required 19 metres. The large windows in the faces of the lantern and in its base allow light to flood onto the main altar stands immediately below it, replacing the monk's choir stalls which stood there originally.

Scarf joints to increase the lengths of timbers took a number of progressively developing forms and have been used to assist in dating structures. The forms ranged from marrying and pegging halved ends to a variety of splayed, notched and pegged linkages. Framed structures such as roofs required the joining of timber members, commonly with mortise-and-tenon joints secured with wooden pegs.

Medieval ridged timber roofs usually conformed to one of four basic forms (Brandon), depending on the means adopted to eliminate or reduce the lateral thrust on the supporting walls arising from the considerable weight of the lead, tile or slate covering, or external loading such as snow or wind, bearing on the sloping rafters: (i) roofs with tie beams; (ii) roofs with trussed rafters; (iii) roofs framed with hammer-beams and braces; (iv) roofs with braces or collars and braces. For timber roofing over a stone vaulted ceiling or wooden ceiling, simple forms such as (i) or (ii) could be readily adopted, as potentially unsightly features such as tie beams or other truss members were not exposed to view. Quite often, however, these roof forms were deliberately exposed to view, with tie beams moulded to become an attractive feature.

A thirteenth-century sketch by French engineer-architect Villard de Honnecourt shows a simple configuration of a hammer beam roof, a method devised to eliminate horizontal ties across the full span at the top of wall level. The simple system illustrated by Villard would have helped against sagging of the rafters and if the joints between the mid-height horizontal member and the rafters had been efficient in tension, would have reduced to some extent the outward horizontal forces exerted by the roof on the tops of the walls.

English medieval builders made extensive use of the hammer beam principle, often with spectacular visual impact, none more so than the roof of Westminster Hall in London. The original hall was built at the end of the eleventh century to plan dimensions of 73 metres by 21 metres, and the division of the building into three aisles longitudinally allowed the roof to be supported at two intermediate points across the spans. Reconstruction of the hall, started under Richard II in 1394 and completed in 1402, consisted of raising the side walls, and roofing the hall in a single span of 21 metres without intermediate supports. The task of accomplishing this fell to Richard II's master carpenter, Hugh Herland. He produced a design of bewildering complexity that has taxed the minds of leading structural engineers ever since, the most complete analysis having been made by Heyman and published in the *Proceedings of the Institution of Civil Engineers* in 1967.

With a slope of 55° (approximately a 3:4:5), triangle rafter lengths are about 18 metres, which meant that Hugh Herland could not fashion these from single lengths of timber, the maximum available lengths of which measured about 12 metres. This restriction clearly had an important influence on the design. Using a conventional elastic analysis and satisfying equilibrium conditions, Heyman showed that under dead weight alone the bulk of the roof loading is carried through the rafters to the tops of the walls, with substantial compression forces also in the crown post and collar beams, but with relatively small or negligible loads in all other members. The resulting outward thrusts of 12.6 tonnnes at the tops of the walls are pointed out by Heyman to be comparable in magnitude to the thrust of a stone vault over a typical French Gothic cathedral nave. The main arch ribs do make an important contribution, however, to the load carrying capacity of the structure, transferring nearly all the lateral thrust from wind loading to the tops of the walls. Very little of the wind load is carried by the curved braces or other members.

By the sixteenth century engineers understood the importance in truss design of the triangle as the only geometrical form that could not be distorted without changing the length of one of its members. Palladio's Cismone bridge is one of the best known early examples of a major truss exploiting this configuration. Its most common use may well have been in the construction of temporary timber supports or centreing for the erection of masonry arches, such as that employed by Sangallo to support the main nave arches in St Peter's in Rome. He was probably copying techniques in use at least since Roman times.

Some of the features incorporated into truss designs and configurations sprang directly from the limitations imposed by the available lengths of timbers and the methods of jointing. This can be seen in the timber truss spanning 20 metres designed by Vasari to support the roof of the Uffizi Gallery in Florence. Each of the sloping rafters is a single piece of timber 11 metres long, but the horizontal member is spliced at mid-span. Vasari produced sketches showing how to scarf tension members by notched overlaps, but these greatly reduced the timber section and thus the strength of the member. Vasari's truss for the Uffizi has a central king post with struts running diagonally from near its lower end to the rafters, above mid-height, with the intention of reducing sagging of the rafters.

Although the interior of Westminster Hall owes much to reconstructions carried out by Henry Yevele, the real highlight is the magnificent hammer beam roof spanning 21 metres built around 1400 by Richard II's master builder Hugh Herland. The view here shows the occasion of James II's Coronation banquet in 1685.

The king post finishes about 100 mm above the bottom chord and is connected to it by an iron strap, thus demonstrating that Vasari understood, correctly, that the king post could not exert any downward loading on the chord and the simple strap simply served the purpose of reducing sagging in the chord member. Under unbalanced roof loading such as wind or snow on one side, some lateral movement of the lower end of the king post must occur with this design and in these circumstances a more positive connection to the chord could have been helpful in limiting possible distortion of the truss.

Scaffolding consumed considerable amounts of timber in the creation of major buildings, mostly in the construction of high walls, but particular features may have called for special temporary or even permanent timbered access, such as that within the lofty spire of Salisbury Cathedral, which acted as staged scaffolding during construction and now provides internal bracing for the thin masonry shell of the structure as well as access for repairs. Special skills had to be exercised where the scaffolding needed to negotiate features such as projections, returns and indentations. In wall construction, the scaffolding would normally have been called upon to support no more than the weight of the workmen, their tools and limited numbers of stones or bricks and quantities of mortar, and usually consisted of slender poles rising vertically from ground level at a suitable distance out from the wall to allow for working platforms of adequate width. Longitudinal horizontal timber members linked the poles at convenient heights and these supported the ends of put-logs inserted into temporary voids left in the masonry and projecting at right-angles to the wall. These, in turn, supported two runs of longitudinal timbers, one near the wall and one just inside the vertical poles, with the platform of timber planking or criss-cross wattle-work resting on these. All timbers were lashed together with stout cords, sometimes tightened by driving in wedges. At high levels such as belfries, towers and spires, the scaffolding had to be wholly supported by the structure itself.

Wooden piles have been used since ancient times in large numbers, and for diverse purposes, including for wharves and weirs, and foundations of buildings and bridges, and many are intact and in place today even from as early as Roman times, a testament to the property of timber, well known to Vitruvius, that it does not perish if it remains permanently below the water table. When the old Pont Notre Dame in Paris, completed in 1507, was replaced in 1853, the new bridge was placed on the same piled foundations. Similarly, the timber piles which had supported the collapsed Campanile of St Mark's in Venice, when examined in 1902, were found to be good enough to support the new structure.

In some cases, at least, the philosophy seems to have been to drive as many as piles as possible so that they touched, or very nearly so, with very little soil remaining between them. The engineer for the Rialto bridge in Venice, Antonio da Ponte, wary of the poor site conditions offered by the soft alluvial soils, in 1588 drove to refusal 6,000 birch-alder piles beneath each abutment, the piles averaging 150 mm in diameter and 3.5 metres in length. Demolition in 1857 of the old Rochester Bridge in Kent, built between 1383 and 1393, revealed that it had been supported by 10,000 piles of elm, each 6 metres long, with a protective barrier or starling of half piles driven around each foundation and the space within filled with chalk. The piers rested on 200 mm of Kentish ragstone, which, in turn, rested on elm boarding covering the tops of the piles.

In his treatise, Alberti set out his recommendations for pile design in marshy ground: that they should be burnt at their ends before driving; the piled width should be twice the width of the wall being supported; their length should never be less than one-eighth the height of the wall and their thickness no less than one-twelfth their length; and finally they should be driven so closely together that there was no room for one more. Furthermore, he counselled that they should be driven, not with heavy blows that could split the timber, but with many light blows. Likening it to driving a small nail into hard wood, he observed that in driving piles 'the continual repetition of gentle strokes wearies and overcomes the greatest hardness and obstinacy of the ground'. Antonio da Ponte seemingly followed this advice in driving his piles for the foundations of the Rialto Bridge, as evidenced by witnesses to the Commission set up to investigate his foundation solution. One witness, a wine merchant, claimed that he watched while it took three hours to drive one of the 3.5 metre long piles and that there existed no possibility that the foundations would fail, which judgement he claimed would be every bit as good as that he could make on a glass of Malvesia wine. Another witness, a salt vendor, attested to the slowness, and hence presumably the soundness, of the work on the basis that he was 'overcome by drowsiness' while watching the pile driving.

Compaction of the ground to strengthen it seems to have often been the primary purpose of piling, rather than relying on the bearing capacity of the piles themselves. This seems to have been the underlying philosophy in Alberti's specification, that the width of piling should be equal to twice the width of the wall carried, that their length should be not less than one-eighth the height of the wall and their diameter not less than one-twelfth their length. Where piles were being driven to be load-carrying, they may perhaps have been driven to virtual refusal by the particular driving equipment being used, or their capacity may have been assessed on their degree of penetration under the blows, based on experience or custom. The first recorded specific use of this method of assessment, still common today, seems to have been in the eighteenth century by Perronet, who suggested that the penetration during the last twenty-five to thirty blows should not exceed 1–2 'Paris lines' or six lines for lesser loaded piles (1 line = 2 mm).

Brick

Although the technology of burnt brick production largely died out in western Europe with the fall of Rome, it survived elsewhere, notably in Byzantine building. One of the world's greatest buildings, the Hagia Sophia, erected by the order of Emperor Justinian in Constantinople as a Christian church, completed in 537 AD, is for the most part a brick building. The huge dome itself is brick, supported by massive marble piers. Islamic builders, too, made extensive use of burnt brick in the construction of their mosques, palaces and fortresses, thus continuing the use of a material long known in the Levant, as well as learning techniques in the use of burnt bricks from the Byzantines, the long-time thorns in the side of Islam. The knowledge of making kiln-baked bricks returned to western Europe through the splendid Islamic structures in Spain epitomised by the Mosque of Córdoba and the Alhambra Palace. Much of the beauty of these structures derives from Moorish skills in making decorative bricks, a technique they learnt from the Persian Sassanians. The Islamic builders achieved a striking visual impact in the horseshoe-shaped arches of the Great Mosque of Córdoba, built in the ninth century by the Umayyad Dynasty, by using alternating voussoirs of red brick and white stone. An interesting structural device in this building employed double arches to gain extra height in supporting the ceiling vaults, something perhaps learnt from studying the Roman aqueduct at Segovia.

Although the Moors in Spain developed their own unique structural and architectural styles, they had behind them over a hundred years of building during the period the Umayyads ruled over the Islamic Empire from Damascus. Most notable was the Great Mosque at Damascus, built between 706 and 715 AD to rival the Christian basilicas of the Byzantine Empire. In fact, the techniques adopted owed something to Byzantium, notably mosaics, and to the Sassanians, from whom the Islamic builders learnt the skills of stucco work (i.e. plaster moulded to give low relief decoration). Texture and colour emerged as important manifestations of Islamic architecture, and strong emotive patterns resulted from the use of burnt brick mosaics, glazed tilework and stucco. With the use of animal or human imagery strictly forbidden by the Koran, decorations derived mainly from foliage forms and the writings of the Koran.

Other outstanding examples of early Islamic brick buildings include the eighth-century Mosque at Kairouan in Tunisia, largely rebuilt between AD 830 and AD 863, and the Ibn Tulun Mosque, built in Cairo in the ninth century to accommodate the troops of its eponymous governor. The main feature of the Kairouan Mosque, its prominence emphasised by the essentially utilitarian style of the building, is an immense hypostyle or columned hall, roofed with a painted wooden ceiling. Built from red fired brick, unusual in Egypt, and faced with carved stucco, the Ibn Tulun mosque excited a great deal of interest in the Islamic world and to a large extent became the archetype of subsequent religious structures of Islam. Features of the building, the third largest mosque in the world, include pointed brick arches, wide corridors on three sides of a courtyard, richly carved ornamental stucco and a prominent minaret with an external spiral staircase. The last echoes that of the famous minaret at Samarra, built earlier in the ninth century and well known to Ibn Tulun himself, who was born in Baghdad, the son of a Turkish slave of Mongol origin, owned by the Caliph al-Ma'mun. He overcame his humble origins to found the short-lived Tulunid Dynasty in Egypt (AD 868–905).

In the period immediately following the fall of Rome, Ravenna in northern Italy became the most important city in western Europe, the Roman Emperor Honorius having made it his capital of the Western Roman Empire in AD 402. After its capture in AD 493 by the Ostrogoth king (and Roman citizen) Theodoric the Great it became the capital of the Ostrogoth Kingdom of Italy. An enlightened man, Theodoric encouraged the arts and crafts to flourish under his leadership, as did the Emperor Justinian when he conquered Italy and brought it under Byzantine influence in AD 538, retaining Ravenna as the local capital. Knowledge of making kiln-fired bricks had survived in Ravenna, as witness a number of surviving structures, such as the Mausoleum of Galla Placidia, built in the fifth century. The bricks used in this structure had dimensions of 400 mm long, 150 mm wide and 100 mm deep, which large size must have made them difficult to fire. By the time the adjacent Byzantine Basilica of San Vitale was completed in AD 547, brick sizes had become considerably reduced, approximately 240 mm by 120 mm. A notable feature of this building was the use of hollow earthenware pots in preference to brick in the construction of the dome, to reduce the weight of the dome and thus lessen the vertical and, in particular, lateral thrusts on the supporting structure. The Byzantine builders made much imaginative use of burnt brick, in some cases alternating it with stone to give striped wall patterns, and also in arch rings, a technique used to great effect by Umayyad builders in the Great Mosque in Córdoba. Islamic and Byzantine builders were not slow to copy each other's techniques, to the benefit of both cultures.

Romanesque building with the Byzantine influence and strong dependence on brick spread throughout northern Italy and across the Alps in the eleventh century to France, then to the Low Countries and eventually to Britain in the thirteenth century. Although never a Byzantine city, Venice nevertheless had close contacts with Constantinople and with Ravenna, and Venetian builders consequently absorbed many Byzantine practices, as shown, for example, in the eleventh-century rebuilding of the Basilica of St Mark. Close trading relations existed in the thirteenth century between Venice and Flanders, the latter having established itself, through production and export of woollen cloth, as the leading industrial area in Europe at that time. Large amounts of the raw wool came from England. Bustling cities in Flanders, notably Bruges, Ghent and Leuven, grew rapidly, making demands on building skills and know-how. The flow of arts and crafts followed hard on the heels of commercial trading, and brickmakers, together with other specialist workers, made their way to these thriving centres in Flanders where their skills were in demand.

Brick sizes at that time, in length and width, varied by an inch or two, but on average corresponded closely to sizes today, i.e. 9 inches (230 mm) by 4.5 inches (115 mm). Thicknesses were much less than today's standards, ranging from 2 inches (50 mm) to 2.5 inches (63 mm).

The knowledge of brick making and brick construction methods spread to many parts of Europe in medieval times, largely through widespread contacts set up by the Hanseatic League. Impressive brick churches of this period can still be seen in a number of these former Hanseatic towns, such as the Marienkirche in Lübeck. Another striking example is the massive Albi Cathedral in the south-west of France, which has claims to be the largest brick building in the world. Although the Catholic Church, with political backing, had, by 1255 and after nearly fifty years of military campaigning, finally suppressed the populist Cathar movement, seen as heretical, nevertheless wary of other possible threats in the future, it deliberately had the cathedral built from its commencement in 1272 as a fortress that could shelter the entire population of Albi, an erstwhile Cathar stronghold. The cathedral's solid brick walls, over a metre thick, rise to some 30 metres in height, strengthened by inward projecting buttresses walling interior chapels. The 89 metre long nave has no aisles and has a roof span of 18 metres. At the west end a massive belfry tower, almost as wide as the nave and with 5 metre diameter corner turrets, rises to a height of 78 metres. Construction of the cathedral continued through the fourteenth century to near completion by 1383, but with some upper storey additions being added over the next century.

Albi Cathedral, built mostly in the thirteenth and fourteenth centuries as a fortress in the wake of the Albigensian wars, has claims to be the largest brick building in the world.

Not surprisingly, some of the earliest examples in England of the use of Flemish bricks are in East Anglia, source of much of the wool that flowed into Flanders. Wealthy merchants used imported Flemish bricks for construction of their manors and churches, each vying to outdo their neighbour. Construction of the Tower of London consumed large quantities of Flemish bricks imported in 1278. Having been perhaps the major port through which Flemish bricks came into England, Hull in the fourteenth century became a major centre for the manufacture of bricks, with many building craftsmen from Flanders settling in Hull and other east coast areas.

From the fifteenth century onwards, the use of brickwork, both as a basic structural material and an ornamental material, became well established in England, leading to outstanding structures such as Hurstmonceux Castle in Sussex, built in the first half of the fifteenth century and now occupied by the Royal Observatory. Built by Sir Robert de Fiennes, Treasurer to the household of Henry VI, its turrets and machiolations betray his earlier career as a soldier: he fought for Henry V at Agincourt. Apart from its chapel, Eton College, founded by Henry VI, was built entirely of brick, a specially constructed kiln supplying some two and a half million bricks for the purpose between 1442 and 1452. In Cambridge, several newly founded colleges were built in brick before 1500, notably Jesus and Queens, the First Court of which, still perhaps the prettiest small court in Cambridge, has survived until today virtually unrenovated. An impressive example of later Elizabethan brickwork can be seen in St John's Second Court, constructed in 1602.

Masonry

Rock strata, which provide the stone for building, fall into three basic categories: igneous, sedimentary and metamorphic. Igneous rocks derive directly from the molten lava underlying the Earth's crust, through which it has pushed up and solidified below the surface or, through volcanic activity, flowed out onto the surface. Lava not reaching the surface (intrusive) cools slowly, allowing individual crystals to form, producing a coarse grained rock such as granite. Lava flowing onto the surface (extrusive) cools rapidly, producing a fine grained rock such as basalt. Historically, the two most common sedimentary rocks used for building have been sandstone and limestone. Sandstones are formed from sand grains (usually durable quartz grains derived from weathering of igneous rocks), bound together by cementing materials such as siliceous matter, calcium carbonate or a ferruginous compound. Limestones are formed from the precipitation in seawater of fine calcareous grains, or by the accumulation on the seabed of shells and skeletons of marine organisms. Igneous and sedimentary rocks can be altered by great heat and pressure to form metamorphic rocks; one of which, marble, formed by the alteration of limestone, found widespread use as a building material in the ancient world.

Limestones and sandstones have been the most widely used historically because of their widespread occurrence and because they can be readily quarried and shaped. A sedimentary stone, notably limestone, laid down slowly over a long period of time with uniform texture and free of bedding planes, and thus not prone to splitting into layers when worked, is referred to as freestone. Harder rocks such as granite and marble have found greatest use for facings and decorative features, or where high strength was at a premium, such as in columns and lintels.

The selection of a stone for building use throughout the pre-industrial age depended on availability, but could also be influenced in no small measure by its location relative to the intended building or structure and thus the means by which it had to be transported to the site. Heavy stone blocks could be transported much more easily on waterborne craft than on wheeled carts drawn by oxen over several miles of often barely formed, muddy or rutted roads. For this reason, and for its excellent characteristics, Cretaceous limestone quarried near Caen in north-western France was used in the construction of several important buildings in London and south-eastern England, despite having to descend the River Orne to the coast, cross the Channel and then up the tidal Thames to London or, for both the cathedral and the castle in Norwich, up the River Wensum to that city. Canterbury Cathedral on the River Stour, only 6 miles from the coast, also consists largely of Caen stone. Initially built to a Norman design and dedicated in 1077, a disastrous fire in 1174 destroyed much of the structure, leading to its reconstruction in the Gothic style with its flying buttresses, prominently pointed arches and ribbed vaulting. In the capital, Caen stone was used for corners and special features in the construction the Tower of London, consuming for this purpose, in year 1278 alone, seventy-five shiploads totalling over 89,000 parpayns – stones with two trimmed faces – at a cost of £332 2s. As well as the relative ease of transporting it, the attractions of Caen freestone for the builders of these structures included its light creamy-yellow colour, its uniform texture free of bedding planes and shells or fossils and its relative softness on quarrying, which made it easy to work and shape. In the buildings themselves, it gradually hardened over a period of time, making it well able to carry any loads imposed on it.

The bulk of the Tower of London consists of Kentish ragstone, a less tractable sandy limestone with round grains of sand and shelly fossils, often used for the filling of walls, with facings of softer dressed stone, in view of its hardness and the consequent difficulty of dressing it. In 1278 304 shiploads of the stone, costing £107 14s 10½d, were delivered to the Tower of London site. Its widespread usage included the bulk of such prominent structures as London Bridge and Dover Castle, the latter receiving, in 1226 alone, 350 boatloads at 10d a load.

Other stones commonly used in England with the advantage of easy quarry access to water transportation included limestones from Barnack and Portland. The coarse grained, shelly limestone blocks won from the Barnack quarries near Peterborough were trundled a short distance overland to the Welland River, whence they could access the Nene and other rivers, leading to its widespread use, including the construction of Ely Cathedral in the Fens. Portland stone, a fine grained limestone consisting of shells and oolites, came from quarries on the Isle of Portland, actually a small rocky peninsula attached to the Dorset coast with consequent direct access to the sea. Inigo Jones favoured its use for reconstruction and refacing work on Old St Paul's Cathedral in London 1633–43, and had a new pier built on the south of the island to which stone was brought on low timber trolleys with two or four solid wooden wheels, drawn by horses or mules along a winding road traversing the cliffs from the main quarry, the King's Quarry on demeyne land. But the subsequent popularity of this stone stemmed mostly from its extensive use by Christopher Wren in his widespread reconstruction work in London following the great fire of 1666.

Wren began the reconstruction of St Paul's Cathedral using Portland stone from the burnt-out Old St Paul's, allowing the Church Commissioners time to prepare and submit to Charles II a petition to import stones from the King's Portland quarries, which the monarch readily agreed to on favourable terms. Wren and Oliver, his Surveyor of Works, immediately set about listing the stones required for various parts of the work, updating these lists at intervals as requested by the master masons. The Isle of Portland soon became a scene of great activity, with the overhaul of access roads, piers and cranes, and the re-opening and extending of the King's Quarry, which had been dormant for many years, requiring the removal by picks and shovels of up to 6 metres of soil and rubbish overburden and blasting away over 4 metres of unusable, poor quality rock to expose the 3 metres or so thick layer of high quality Portland stone. The blocks were then marked out to the required size and then freed by driving iron wedges into 150-mm-deep slots cut into the rock along the etched lines. After the blocks had been scappled by axe (rough dressing), trolleys conveyed them to the pier, where cranes loaded them onto the ships. The most difficult part in conveying the stones to the cathedral site came at the end of their journey, as the ships, unable to pass through the narrow arches, had to discharge their loads downstream of London Bridge. The heavy blocks had then to be hauled on horse-drawn trolleys through busy, narrow, poorly paved streets, including up the steep St Bennet's hill, to their final destination. Here the blocks were sawn to templated shapes and sizes, and finally dressed and smoothed using chisels and other special tools, ready for inclusion in the growing structure.

Although recognised as an excellent freestone for building, the expense of overland transportation greatly restricted the use of the honey-coloured Bath stone, an oolitic limestone, during the middle ages, other than within a few miles of its quarrying areas in Somerset. Usage of the stone became more widespread after the completion of the Bristol Avon navigation scheme at Bath in 1727 and more widespread still after completion in 1814 of the 57-mile-long canal linking the Bristol Avon navigation to the Kennet River navigation. Its usage now extended as far as London. Another specialist stone used extensively throughout the medieval period in cathedrals in the south of England was so-called Purbeck Marble, in fact not a metamorphic stone like true marble but a sedimentary limestone consisting of the closely packed shells of fresh-water snails. Sandwiched between layers of softer, marine deposited clays and mudstones, it came from quarries excavated into a peninsula on the Dorset coast and consequently lent itself to over-water transportation. Capable of being finished to a high polish like true marble, it found particular use in applications such as columns, flooring and slab panels.

The strength of any building stone, and hence its suitability for specific usages, depends not only on its basic composition, but also on flaws within it such as bedding planes, joints and fissures, which are not always apparent on visual inspection. The types of stones used for building have high intrinsic compressive strengths, making them very suitable for walls and mass structures, but even the hardest can only be relied upon to exhibit limited tensile strength, with any flaws or incipient cracks further decreasing this. Notwithstanding this, tensile stresses can and do occur in masonry structures in certain circumstances, notably where stone beams or lintels are used. Deepening a beam will in theory increase its strength, but this advantage may be cancelled out by its own increased weight and the greater possibility of cracks or other flaws occurring in the section. The formation of a vertical crack through the depth of a deep lintel, while unsightly, may not lead to its collapse as it may then act essentially as an arch, but this imposes very high horizontal compressive stresses at the top of the crack, balanced by high thrusts against adjoining lintels. The limiting span for a limestone lintel is about 3 metres, but may be up to 8 metres for harder stone. While this inherent weakness and unreliability of stone beams and slabs to span distances or cover open spaces did not entirely deter builders, they resorted more usually and logically to the arch form, following the lead of their ancient and classical forebears, many of whose structures would have been familiar to them in intact or partly ruined state. In the latter case, arch rings were sometimes to be seen surviving when much of the remainder of the structure had crumbled. These ancient structures often became, in effect, quarries from which the new builders derived much of their masonry in the form of blocks, columns and the like.

Masonry arches, vaults and domes were usually constructed of fitted stone blocks, which exploited the compressive strength of the stone, in some cases achieving huge dimensions. A notable exception to the use of stone blocks being the monolithic 11 metre diameter dome of the church of Theodoric in Ravenna (*c.* AD 530), fashioned from a single piece of Istrian limestone. Arches took either a smooth curved form or a pointed form, both of which leant themselves to a variety of specific shapes, the adopted form or shape governed by fashion, precedence or expectation.

Mortar

Historically, many different substances have been pressed into service as mortar, including mud or moistened earth, most commonly for mud brick structures and sometimes mixed with chopped straw or reeds, gypsum, pitch, a mixture of sand and lime or sand and cement. Medieval and post medieval builders had to rely on sand and lime mortar up to the nineteenth century, when the invention of Portland cement, first patented by Joseph Aspdin in 1824, gave, by its inclusion in the mix, the option of a harder and stronger, but more brittle, mortar. While the role of mortar has always been primarily to bind together the bricks or masonry of a structure, it has, particularly in masonry structures, often served to even out the pressure between blocks and thus eliminate high localised contact stresses. It is likely that the ancient Egyptians used gypsum mortar for a third purpose, namely as a lubricant, making it easier to slide the blocks, weighing several tons, into position.

The quicklime used for mortar was obtained by crushing chalk or limestone to a powder and then baking it in a wood or coal fired kiln, usually at the quarry site or at a commercial site such as Sea Coal Lane, also known as Limeburners Lane, at Ludgate, just outside the walls of London. The smoke nuisance drew many complaints from city dwellers. At the building site, water added to the quicklime converted it to calcium hydroxide or 'slaked' lime, which, when added to sand, in sand/lime proportions commonly between 1.5 and 3, produced a mortar putty. It was considered good practice to leave the mortar for a day or two before giving it a final vigorous mix before use. In place, it hardened very slowly, over years, as it absorbed carbon dioxide from the air, converting the calcium hydroxide back to calcium carbonate. The slow hardening had the favourable outcome of allowing masonry structures, during and after construction, to adjust to modest movements arising from such as foundation settlements without unsightly cracking.

Iron

Its widespread use in tools, weapons and farming implements, and in adminicular items in the building industry, made iron by far the most important metal produced throughout the pre-industrial age. Methods of reducing iron from its ore differed very little in medieval times from those that had been in use for the previous 2,000 years and more. The ore had to be broken down into chunks before introducing it into the furnace, which often consisted of no more than a depression in the ground with a clay lining and dome. This gradually gave way to an above-ground enclosing structure of clay and stone with openings for gas exit and admission of the ore, and an aperture near the base to admit a draft from bellows and extraction of the 'bloom' of iron. Soft and glowing, but not molten, the bloom had to be pummelled on a hard, flat surface to rid it of slag and prepare it into shapes suitable for the smith. This process of producing wrought iron, which typically required 12 lb of charcoal for every 1 lb of iron produced, contributed to the denuding of swathes of forest across Europe. Steel could be produced by the complete expulsion of slag and the cementation process of further repeated hammerings and heatings in direct contact with charcoal, to raise the carbon content from 0.05 per cent or less in wrought iron to between 0.5 per cent and 1.5 per cent for steel and, by further heating and quenching, the steel could be given an increased hardness. The advent of very tall furnaces (reaching 9 metres in height by the seventeenth century), with stronger blasts from water-powered bellows, and using harder charcoal, made it possible to melt the iron, thus enabling it to absorb larger amounts of carbon and produce the harder and more brittle cast iron with carbon contents of 3 per cent to 4 per cent. The molten iron settled at the base of the furnace with the lighter molten slag, produced from melting of the limestone used as a flux, floating on top of it. Periodically the slag was first drawn off, followed by the iron, which flowed into a channel with secondary troughs leading off it, giving rise to the term 'pig-iron', as it seemed to resemble a sow feeding her piglets. Application of this technique, discovered in the thirteenth century, developed slowly, taking another two hundred years or so for cast iron to be produced on a large scale. Although never challenging charcoal in iron production, coal did find some limited use in the early stages of smelting, but could not be used in later stages because of its contamination of the iron with sulphur and phosphorus.

Iron production spread rapidly throughout Europe, largely through the Cistercians, who showed great skills in acquiring ore-bearing lands. Burgundy and Champagne became major Cistercian iron and steel producing areas in France, their operations in Champagne centring on the Abbey of Clairvaux, where the monks went on acquiring ore deposits and constructing forges from medieval times through to the end of the eighteenth century. They sold off any surplus not needed for their own use. Because of the reliance on water power, the smelting and other forge operations in iron working and production were mostly sited close to rivers. Most of the iron used in England came from the Forest of Dean or the Sussex Weald, but the severe deforestation in these areas led to the centres of production moving to South Wales, Shropshire and the Midlands. Some iron was imported from Spain and Normandy and particularly Sweden, which, with its rich ores and abundance of timber, produced bar iron and steel of very high quality.

The routine use of iron as a structural material had to wait until the eighteenth century, when the coke-smelting of ore became firmly established and played no small part in the subsequent industrialisation of Europe. Nevertheless, iron served the building and construction industries in a large number of ways in pre-industrial Europe. Every village and major construction site had its smith, a respected figure in the community. Many of the basic tools, such as picks, axes, spades, shovels, hammers, rods and saws, were made of, or shod with, iron or steel; iron shoes greatly improved the performance of the horses, widely used for haulage and other heavy duties; iron shoes protected the tips of timber piles driven into the ground. Stores in Calais in 1390 recorded 494,000 nails, evidence of the vast numbers used by carpenters. The invention in the fourteenth century of the draw plate, with a series of successively smaller holes, greatly increased the production of wire, previously forged by hammer. The wire drawer sat on a swing pulled back and forth by a water wheel and crank and on each back swing grabbed the wire with tongs, pulling it through a hole in the draw plate as he swung backwards.

Although not used for primary structural members, iron certainly served in an important capacity in the form of cramps, stays, tie-rods and dowels to provide reinforcement in many important masonry structures. Tie rods were commonly used between the springings of arches in the absence of any external means of absorbing the horizontal thrusts. Brunelleschi inserted iron chains near the base of his dome of Santa Maria del Fiore to aid his chain of sandstone blocks, linked by iron plates, in resisting outward thrust. Wren made similar use of iron chains in St Paul's Cathedral. A mathematical analysis of cracking in the dome of St Peter's in the eighteenth century resulted in its being encircled with additional iron bands to strengthen it. The use of iron for reinforcing could lead to its own problems. At Sainte-Chapelle in Paris iron chains inserted to strengthen the masonry eventually caused it to crack and flake, perhaps as a result of rusting leading to volume increase and expansion of the iron. Knowledgeable builders knew of this problem and boiled the iron in tallow to prevent it.

Tools and Techniques

Throughout the pre-industrial age humans and animals, aided by simple mechanical devices, provided the only practicable available sources of power, together with water power and, in some special circumstances, wind power, to address the many operations involved in construction, such as excavating and quarrying earth and stone, felling timber, shaping stone, transporting building materials to the site, lifting into place and positioning. The lever was the simplest, oldest and most widespread device used to boost muscle power: a 50 kg pull exerted by a man could be converted into a force of half a tonne by a lever with a 10:1 ratio for purposes such as freeing stone blocks in quarry operations, lifting the blocks onto a means of transport to the building site or raising the blocks into their final position in the structure, although for the last application it could be too unwieldy and slow to be the preferred method and may have been employed only in final positioning of the block. The creation of rope from a variety of plant fibres furnished man from earliest times with an invaluable aid in construction, and without it many of the great civil engineering works throughout history would probably not even have been attempted. It made possible the enhancement of muscle power by means of treadmills, pulleys, cranes, capstans and windlasses, as well as providing a means of exerting a direct pull on an object.

Building operations for masonry structures started in the quarries, where the stone blocks had to be prised from the parent rock using picks, chisels or wedges hammered into the rock to release the block, no doubt making full use of natural weaknesses in the rock, such as joints and bedding planes. Levers could then have been used to manoeuvre the block into position for preliminary shaping, to reduce to a minimum the volume of rock which had to be transported to the site. Shaping would have been according to templates supplied from the building site, and carried out by scappling with sharpened steel-edged iron tools, such as chisels or axes, and possibly by the employment of two-handled crosscut saws lubricated with water. In some cases final shaping may have been done at the quarry, particularly if the block was to be left for 'seasoning' to harden it before transporting it to the building site.

In seeking a suitable quarry for the supply of stone, both the quality of the available stone and its transportation to the building site had to be considered; the latter could easily exceed the former by a substantial amount. A nearby quarry might be used to supply the bulk of the stone for the building, with better quality stone being brought in from greater distances for specific usages, such as architraves, mouldings, voussoirs, plinths and columns. Various arrangements existed between the quarries and those commissioning the buildings, in some cases the latter, such as the Crown or a great monastery, owning the quarry outright, in other cases leasing the quarry using their own appointed quarrymen, to buying in stone directly or through middlemen from quarries independently owned and worked.

Transportation of light blocks could be by wheeled handcart or borne on two parallel poles with cross-members supporting the block, carried by two men, one at the front and another at the back. Larger blocks required wheeled carts or wagons, necessarily heavy in construction to support the weight of the stone and to negotiate the rough, often potholed or waterlogged roads of the time. No doubt they still broke down frequently and required constant maintenance. Oxen and steer were commonly used to draw these carts and on occasion horses, the hauling power of the last greatly enhanced by the introduction of the iron horseshoe and by an improvement in harnesses, the throat girth of antiquity having been replaced by a collar resting on the shoulders of the animal to which the pulling ropes were attached. Waterborne transportation was by river barge or coastal shipping.

The much higher cost of land transportation than carriage by water placed a premium on quarries, and possibly even building sites, situated close to rivers, canals or the coast. The excavation of a short linking length of canal to a river or the coast, particularly for a large, important building, could also prove economical. For distances greater than some eight miles or even less, the cost of land transportation exceeded the cost of the stone at the quarry. In 1237 'two hundred of freestone' costing 3s at the quarry near Bath in England cost a further 22s on delivery to Marlborough, a distance of some 32 miles; the high quality and uniformity of this Oolitic limestone, which could be sawn in any direction, presumably justifying this addition to the cost. This contrasts with the purchase of a similar high quality limestone shipped from Caen in France to London in 1429 for London Bridge, costing 2s 6d at the quarry and a further 5s for carriage.

Lifting devices and methods at the quarries took a variety of forms. At the quarry, for blocks too heavy to be manhandled by unassisted human power, levers could have been used for small lifts, but the placing of the blocks onto a cart or floating craft would have demanded the use of some form of crane or other hoisting device. At the site lighter stone blocks and features could have been carried, by hand, up ladders, but heavier blocks would have required some form of hoisting device. Two common devices used long timber poles, commonly 4 metres or more in length, set up as a tripod or gin lashed together at the top with an attached pulley or two poles set up as spread shear legs, again lashed together at the top with an attached pulley. A rope passing over the pulley and fastened to the block could be hauled on by hand or by windlass anchored in some way at a suitable height, possibly to the poles themselves. The gin had the disadvantage that it could only lift from a fixed position, without being moved bodily, and offered little scope for lateral movement of the block at its designted height. Shear legs, held by a guy rope from the apex to a secure anchorage, could be tilted to allow the block to be picked up from a limited range of positions in a stack and also gave greater flexibility in placing a stone into its allocated position. Shear legs could also be set up on a part of the structure itself, such as a flat roof or building projections, or on a specially provided platform resting on part of the structure.

Early forms of cranes consisted of vertical posts with a horizontal arm at the top, either in the form of a T or inverted L, with pulley wheels at each end of the arm over which the hauling rope passed. This configuration, a primitive forerunner of the modern tower crane, had the advantage that the assembly could be swivelled, allowing the object being lifted to be placed where required. A structure resembling a modern tower crane figures among the many sketches of Leonardo da Vinci. The later medieval and middle ages saw the evolvement of much more powerful cranes, with the hauling ropes from the operating winch passing over pulleys at each end of a heavy inclined beam or frame, capable of being elevated to different working heights, resting on a heavy timber tripod or framed structure, on which it could swivel.

The completion of the external structure of a great cathedral allowed the installation of windlasses in the roof with a hauling rope wound around a drum of much smaller diameter than its attached great winding wheel, thus giving a large mechanical advantage. These allowed the hauling up of heavy objects, such as large timbers, vaulting ribs, keystones and slabs, arch centrings and bells. The great wheel could be fitted with spokes around its periphery and turned by hand or, as in some cases, it could be actually be in the form of a treadmill operated by one or more men. A number of examples of great wheels are still in existence, including Mont Saint-Michel in northern France, and Tewkesbury Abbey and Salisbury and Peterborough Cathedrals in England. The great wheel in Peterborough Cathedral has a diameter of 3.66 metres. As seen in the sixteenth-century book of Ramelli, the possibilities of using gearing in lifting devices was well understood and Leonardo in one of his sketches shows an exploded sketch (an innovation in itself) of a device in which a drum with a rope wound round it could be used to lift objects by the to and fro operation of a lever, with the object being lifted by both of the motions.

Pile driving seems to have been a common practice throughout the pre-industrial age, and early pile drivers may have consisted simply of a tripod of timber poles lashed together at the top with a suspended pulley, over which a rope passed, fixed to the drop hammer at one end and operated manually at the other. The hammer would have been hauled to a chosen height then the rope released to run freely over the pulley, allowing the hammer to drop onto the head of the pile. A much more sophisticated pile driver illustrated by the fifteenth-century Italian engineer Francesco di Giorgio shows the drop hammer sliding between, and thus confined laterally by, vertical frame members (leaders), and the rope operated by a winch. The assembly incorporated an ingenious feature in an inclined hook attached to the rope to hold up the hammer, so that on reaching the top of the drop height, the free end of the hook engaged a fixed bracket and released the weight. Pile drivers today differ little in principal from that illustrated by di Giorgio, but enjoy the benefit of powered drives. The thirteenth-century French engineer Villard de Honnecourt sketched a hand operated device for sawing off the tops of piles under water, which he may have noted in operation, if indeed it existed, or it may have been his own invention. A saw blade, fixed between the lower ends of two long poles held apart by a framed structure, could be slid back and forth, constrained within a slot in a fixed timber frame. A rope with an attached heavy weight passing over a pulley wheel and secured at its other end to the frame separating the two poles kept the blade held against the pile.

Muslim engineers derived water for irrigation and domestic use from underground where surface sources were not available. Vertical wells provided the simplest solution, but produced only a slow intermittent extraction as the water had to be raised by rope and bucket, hauled up either directly by hand or by winding the rope around a drum and raising the bucket by windlass. The shaduf, a device dating from ancient times, also found use in the Islamic period for raising water from rivers or wells. It consisted of a long pole balanced on a short cross-beam, which acted as a fulcrum and spanned between the tops of two high vertical posts, one end of the pole having a bucket suspended from it by a rope with a counterweight at the other end of the pole, often just a wodge of clay. By hauling on the rope, the operator could lower the bucket into the well then, with minimum effort, raise the filled vessel and discharge the water into a reservoir or channel. A larger and more continuous supply could be extracted from wells using the saqqiya, which consisted of a horizontal wooden lantern gear wheel (parallel wooden discs connected by vertical wooden dowels spaced around their periphery), driven by animal power, engaging with horizontal pegs projecting from the periphery of a vertical wooden gear wheel, the latter attached to a drum carrying a looped chain of pots extending down into the well. To raise the water the animal, usually a donkey or mule, or even a camel, walked around a circular path, tethered to the end of a bar which slotted at its other end into an extension of the vertical axis of the horizontal gear wheel.

Islamic engineers made wide use of the noria, a large, water driven wheel used to raise water to a high level, and already known in Roman times. The wooden wheel, up to about 20 metres in diameter, turned on an iron or wooden axle, and the outer rim had fixed to it up to 120 compartments that picked up the water and delivered it to a head tank, whence it discharged into a high level aqueduct. The undershot wheel was driven by water in a fast flowing stream impacting on wooden paddles radiating from the rim and fixed between each pair of compartments. In some cases, earthenware pots were strapped to the rim of the wheel in place of the compartments.

In 1206 Al-Jazari, an engineer in the service of the Turkish Artuqid family, published his book on *Ingenious Mechanical Devices* that included a number of proposed animal and water driven geared mechanisms for raising water. His sketches also include methods of converting continuous input motion by animals into intermittent motion for operating water raising swapes and also show the first known introduction of a crank into a machine, although hand operated cranks had long been in use. The first two

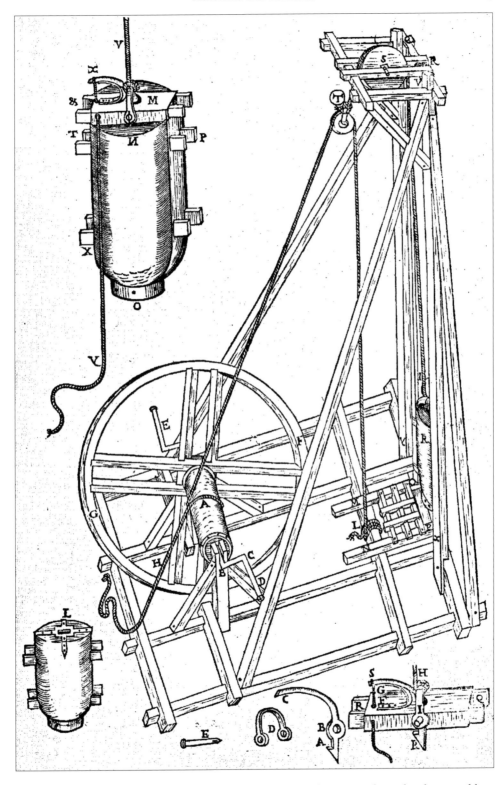

A sixteenth-century pile driver operated by a windlass, and with a trigger device hand operated by a cord to release the hammer at the top of its lift.

machines he described relied on animal power, but the third was a water-driven version of the saqqiya, operating as something of a showpiece beside an ornamental lake. Water from the lake flowed into a tank with a hole in its base, through which it descended onto a scoop wheel turning on a horizontal axle, located in a tank below the water-tank, which also contained the gearing converting the motion to rotate a vertical shaft. The top end of this shaft had further gearing driving a chain of pots, which extracted water continuously from the upper water tank. The vertical shaft also rotated a wooden model of a cow, its feet barely clear of a horizontal platform, giving the false impression that it was driving the gearing.

Wind and water offered the only practicable sources of power, other than human or animal effort, until the harnessing of steam in the eighteenth century, and both of these sources found widespread use throughout Europe for a range of applications, in contrast to their use by the Romans and Greece almost solely for the grinding of corn. Machines had to be devised and constructed to exploit these sources and these came primarily in the form of windmills and watermills, the former commonly used in land drainage and grinding corn, and the latter used for many different and varied applications. In flat landscapes wind provided the only practicable source of power other than that afforded by human or animal muscle.

In order to drain low-lying, inundated land to make it suitable for cultivation or habitation, the windmill had to lift the water from the field and discharge it into a purposely constructed drainage channel or lode, or into a river. The rotating sails of the windmill drove, through gearing mechanism, a vertical scoop wheel, comprising a wooden, later an iron, wheel, with wooden or leather flaps projecting beyond its perimeter and usually slanted back about 30° to the radial direction. It revolved in a shaped trough from which it lifted the water into a higher trough, with the lift limited to about one-third the diameter of the wheel, above which water would simply have flowed back into the lower trough. Thus, a wheel 4.5 metres in diameter was needed to raise the water by 1.5 metres. For high lifts, two or three windmills or possibly more would have been used in succession to raise the water to the required height.

The two essentials of windmill design were the capability of turning the sails to face the wind and an internal mechanism to transfer the rotary motion provided by the sails at the top of the tower to drive the scoop wheel at or near ground level. In early windmills, dating from ancient times primarily to grind corn, the entire wooden structure of the mill, which rested on a substructure fixed into the ground, had to be turned bodily into the wind, requiring great effort; and for water-raising suffered the added and more serious shortcoming that it could not drive a fixed scoop wheel. The invention of the tower mill in the fifteenth century, with a revolving cap, solved both these problems and, in addition, allowed the body of the tower to be constructed in much more solid and durable brick or stone. A vertical shaft passing down through a hollow post supporting the body of the mill transmitted the drive from the turning sails to the scoop wheel below.

Water provided the only reliable and consistent source of power for many purposes, other than humans and other animals, throughout the pre-industrial age; its provision of power derived either from its weight and elevation, i.e. potential energy, or from the

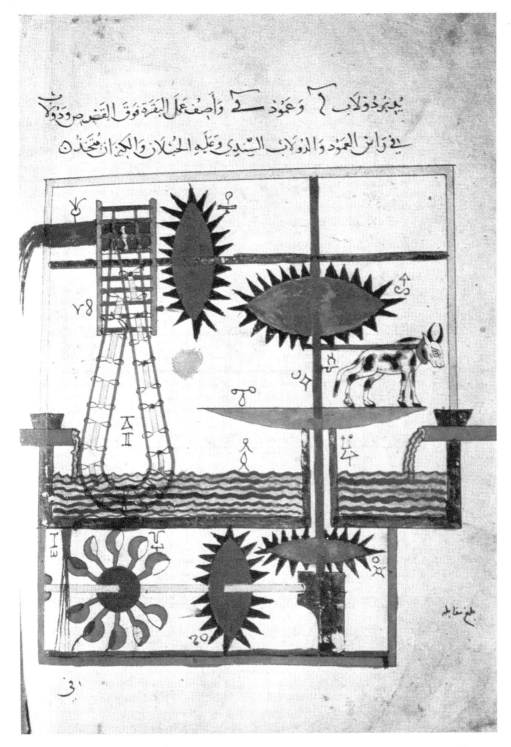

Al Jasari's machine to raise water from a lake for ornamental purposes using a water-powered scoop wheel.

kinetic energy of flowing water. Consequently, waterwheels needed to be sited close to or within hilly or sloping land, or in mountainous country. Confined almost solely to the grinding of corn in the ancient world, waterwheel usage exploded in medieval and subsequent times to embrace a wide variety of applications as illustrated, for example, in the sixteenth-century treatises by Ramelli and Agricola. In one area alone, iron production, water power found many applications, including stamping mills to break up the ore into manageable pieces for the furnaces, trip hammers to forge the blooms, rolling mills and operating bellows for the furnaces.

Waterwheels, constructed from wood or iron or often some combination of these, took a number of forms within the limitations of either rotating in a horizontal plane about a vertical axis, or rotating in a vertical plane about a horizontal axis. The former demanded for the most part better water control, but had the considerable advantage of not requiring any gearing or machinery in the grinding of corn, as the upper end of the driven vertical shaft passed through the lower fixed millstone and engaged with the upper runner stone. The most basic form consisted of directing a controlled flow of water from a head race and down a steep wooden chute onto plane flaps or vanes projecting radially from the outer edge of the wheel. This required the water to be supplied to the site at an elevation above that of the wheel. More sophisticated forms, which benefited particularly from high water pressures where available, had curved or spoon shaped vanes with a strong, near horizontal, water jet from a nozzle impacting on these, thus anticipating the Pelton wheel and modern turbines.

The power to drive vertical wheels came, in the case of undershot wheels, from the kinetic energy or impulse of moving water, and for overshot or breast wheels from the weight and potential energy of the water. Undershot wheels had paddles projecting radially from their circumference dipping into fast running water. Maximum efficiency in operation, which rarely exceeded 22 per cent, required the fast running water to be confined within a wooden channel. Other means to increase efficiency included fashioning paddles from double boards set edge to edge at an obtuse angle to give a scoop-like effect and providing a small drop in the tailrace to prevent back up of the tail water, known as back-watering. Wheel speed, and hence output, could be controlled by a moveable vertical hatch or gate upstream of the wheel to control the flow of water. A late development by French engineer Poncelet around 1824 saw an increase in efficiency achieved by the introduction of a carefully controlled inclined hatch and improved paddle designs.

Overshot wheels operated by having buckets around the outer rim receiving water discharged onto the top of the wheel, requiring the water to be diverted from higher up the river and brought to the mill site along a gently sloping channel or leat. For maximum efficiency, the buckets had to be shaped to carry the water for most of their descent but to fully discharge it close to or at the bottom of their descent. Effecting the diversion usually required the construction of a weir across the river and, not uncommonly, a holding pond, or mill pond to ensure the water supply. With their much higher efficiency, overshot wheels needed only about a quarter the volume of the water needed for undershot wheels. In very steep or mountainous country, even higher efficiency could be achieved by having the head race at a height sufficient to allow it to have an angle of

A sketch by Leonardo showing the utilisation of a water wheel to lift water by means of Archimedean screws to the top of a tower.

around 30° to the horizontal at its approach to the wheel and, by impacting the wheel at about one o'clock, delivered both a kinetic impulse as well as potential energy. In the conventional overshot wheel the top rotated away from the head race, but occasionally millers favoured a contrary rotation, known as a pitch-back wheel, requiring a stop board at the end of the tail race and for the water to fall onto the wheel through a slot in the floor.

The breast or breast shot wheel provided a solution to exploiting the potential energy of water where a sufficient elevation of the water channel could not be achieved for an overshot wheel. These wheels had cups or buckets similar to those for overshot wheels, but the water flow was directed from a chute into these at the front of the wheel at about mid-height, in some cases slightly above or below mid-height, and consequently revolved in a direction opposite to that of the conventional overshot wheel. The provision of a vertical hatch in the chute allowed control of the amount of water directed onto the wheel. Particular attention had to be paid to keep the system clear of debris, which could lodge between the buckets and the end of the chute and stop the wheel or even badly damage it. An efficiency of up to about 55 per cent could be attained with a good installation, well below that attainable with an overshot wheel, but well above that for an undershot wheel.

In building and infrastructure works, waterwheel use was confined mainly to water raising and operating sawmills. Leonardo da Vinci sketched an arrangement consisting of two Archimedian screws linked to a waterwheel to raise water to the top of a high tower. The installation of a water wheel often required the construction of a special channel feeding from a river, the design and construction of which could be an engineering challenge in itself, as the channel had to deliver the water with an adequate velocity to drive an undershot wheel or deliver it at an adequate height to drive an overshot wheel. In some cases, advantage was taken of water rushing between masonry bridge piers to drive waterwheels to raise water.

In order to provide timber in its required shape or structural form, such as planking, squared post or strut, or rectangular beam, if done by hand, it had to be hewn by axe or adze, cleft by driven wedges, or sawn, usually by a two-man saw operated over a pit. Mechanical saws driven by water-power were developed, or at least sketched, notably by Villard de Honnecourt in the first half of the thirteenth century and by the Renaissance engineer Francesco di Giorgio in the second half of the fifteenth century and eventually became widespread in use. Villard's sketch is the first known representation of an automatic powered machine performing two motions simultaneously: one a reciprocating motion operating the saw itself, and the other progressively advancing the log. It is unlikely to have been constructed or to have worked if it had been. In Francesco di Giorgio's more sophisticated and more convincing sketch of a hydraulic saw, a crank driven by the rotating waterwheel imparts a reciprocating motion to the saw and at the same time advances the timber piece being sawn by an imaginative assembly of two rods and an oscillating beam, such that the end of one of the rods impinges onto a toothed wheel, giving it a intermittent motion, similar to a clockwork mechanism, and translated by means of an wound rope to the movable carriage on which the timber piece rests.

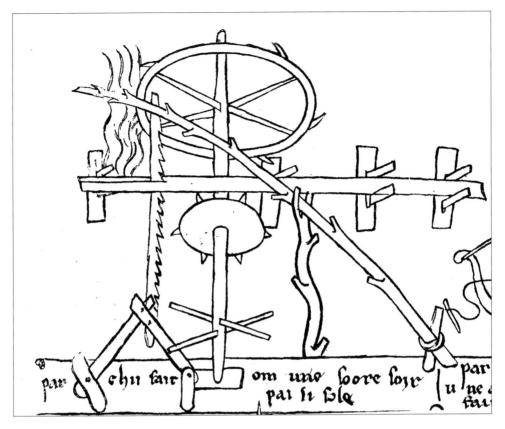

A thirteenth-century sketch by Villard de Honnecourt of a crude water-powered contrivance for sawing planks. It is the first known depiction of two simultaneous motions of sawing and moving the log forward.

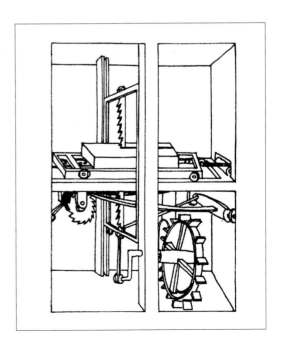

A fifteenth-century drawing by Francesco di Giorgio of a water-powered sawmill.

3

Water Controlled

Domestic and Public Water Supply

Islam inherited the qanat from earlier civilisations such as Persia and Assyria, a method that supplied a continuous flow from underground sources without the need for any mechanical devices, but did need remarkable surveying skills, and an understanding of geology and the distribution of water below ground. A qanat is a gradually sloping tunnel tapping an underground water source in the hills and conveying the water underground to the plains at the foot of the hills, where it discharges into water retention basins or into a system of irrigation channels. The tunnel slope is usually between 1 in 500 and 1 in 1,500. It is constructed by sinking a line of vertical shafts, some 50 metres apart, along its projected line, then connecting the bottoms of the shafts by a continuous tunnel, a not inconsiderable exercise in surveying. The vertical shafts serve as ventilation ducts and also for the removal of spoil from the tunnel excavation. Many of these systems are still in use today, most notably in Iran, where 50,000 km of these underground conduits provide up to three-quarters of all water used in that country, including irrigation water.

According to the Islamic writer Al-Muqaddasi, there were many norias on the river at Ahwaz in Iran, delivering water into aqueducts, along which it flowed into holding cisterns, or into channels taking water to the orchards for irrigation. Another writer, Al-Idrisi, described in 1154 a noria over 40 metres in diameter that lifted water from the River Tagus in Spain and discharged it into an aqueduct supplying the city of Talavera. A fine example of a similar noria can still be seen on the River Orontes at Hama in Syria. Where the stream flow was insufficiently strong, dams were built to raise the head of water to overcome this problem. One such scheme in Khuzistan supplied no less than ten norias, the raised water flowing through channels to 300 villages.

Not surprisingly for a city replacing Rome as the capital of the Roman Empire, the early emperors of Constantinople instigated major works to ensure an adequate water supply. Their engineers had to go to distant Thrace to find springs that could provide an adequate and reliable source of water and they constructed, initially, a channel over 150 km in length in the fourth century in the reign of Emperor Valens, after whom the aqueduct is named, to bring the water to the capital. It took about 30 years to build.

Extensions to the scheme over the next century and a half bringing water from further sources, both by extending the original channel, and by means of tributary channels feeding into the main channel, increased the total length of water channels to over 400 km, with the longest continuous stretch from near modern Vize to Constantinople having a length in excess of 250 km. Although open excavated channels comprised much of the length, the mountainous nature of the country made necessary many kilometres of underground tunnels and the construction of up to 100 arched stone structures (sixty have been positively identified, of which nineteen are substantially intact) totalling several kilometres in length, mostly one tier in height, but with a small number comprising two or three tiers and up to 30 metres and more in height. At its peak the system delivered up to 130,000 cubic metres of water a day to serve the domestic and public requirements of its half a million population. The system would have needed constant maintenance to keep the water flowing, and there is evidence of major works having been carried out in the sixth century in the time of Justinian I, probably to repair earthquake damage. In 626 the Avars, in laying siege to Constantinople, cut the aqueduct and the water supply to the city was not fully restored until 758 by Constantine V.

The city needed large water storage units to tide over periods of drought or siege. These included both private and public storages, numbering perhaps a hundred or more, the public storages comprising both open reservoirs and covered underground cisterns. The three major open reservoirs were the Aspar, 152 metres square and 10 metres deep; the Cukurbostan, 170 metres by 147 metres and 10.5 metres in depth; and the Etius, now used as a football field. Underground cisterns provided huge water storage capacity, some of which, including two of the largest built in the time of Emperor Justinian (527–65), are still in use. One of these, the 'Cistern of a Thousand Columns' has walls nearly 3 metres thick and a height of 18 metres, with a roof of brick vaults supported by sixteen rows of fourteen columns encompassing an area of 3,500 square metres. Even larger is the cathedral-like Yerebatan Sarayi, close to the Hagia Sophia, with its 365 marble columns in twenty-eight rows spaced 4 metres apart, the ornate capitals of the 8 metre high columns supporting a roofing of small brick vaults and embracing an area of nearly 10,000 square metres. Stone columns from older buildings comprise part of its construction.

When Sultan Mehmet conquered Constantinople in 1453, he inherited a water supply system in a reduced state, largely because of reduced demand from a population only a fraction of that of the city at its peak. Much of the long distant aqueduct bringing water from Thrace had been reduced to a state of disrepair by earthquakes, military activity and simple neglect and had consequently been abandoned in favour of bringing water from two sources closer to the city – the Belgrade forest to the north of the city and Halkili to the west of the city – but making use where possible of existing aqueduct structures. The Ottomans repaired and extended these systems, much of the work being carried out in the time of Süleyman the Magnificent under the direction of his engineer Sinan. Masonry dams at various locations along the streams within the forest of Belgrade fed water into the system at distances up to 30 km to the north of the city. Around eight hundred fountains were set up in the city for public use, some highly ornamented and architectural gems.

The underground cistern Yerebatan Sarayi in Istanbul, built in the sixth century, embraces an area of 10,000 square metres, with 365 marble columns supporting a roof of small brick vaults.

Building activity in the time of Justinian is ably recounted by Procopius, who was born in Caesarea in Palestine in the late fifth century and became legal advisor and secretary to Justinian's general, Belisarius. In accompanying Belisarius in his campaigns, he saw much of the Empire that under Justinian, albeit briefly, reached its greatest extent since the time of Constantine. His travels obviously took him to the extreme eastern edge of the Empire, where he describes the construction of a dam to control floodwaters which had breached the walls of Daras, close to the Persian frontier, and caused severe damage within the city. In addition to its flood protection role, the water held behind the dam apparently provided a water supply for the city, as Justinian, according to Procopius, built a great conduit to lead water to every part of the city. Justinian at first sought the advice of his master-builders Anthemius and Isidorus, but, apparently by divine inspiration, chose a scheme put forward by another master-builder, Chryses of Alexandria. Procopius' description of the dam is worth recounting:

At a place about forty feet [13 m] removed from the outer fortifications of the city, between the two cliffs between which the river runs, he [Chryses] constructed a barrier of proper thickness and height. The ends of this he so mortised into each of the two cliffs, that the water of the river could not possibly get by that point, even if it should come down very violently. This structure is called by those skilled in such matters a dam or flood-gate, or whatever else they please. This barrier was not built in a straight line, but was bent into the shape of a crescent, so that the curve, by lying against the current of the river, might be able to offer still more resistance to the force of the stream. And he made sluice-gates in the dam, in both its lower and upper parts, so that when the river suddenly rose in flood, should this happen, it would be forced to collect there, and not go on with its full stream, but discharging through the openings only a small volume of the excess accumulation, would always have to abate its force little by little, and the city-wall would never suffer damage. For the outflow collects in the space which, as I have said, extends for forty feet between the dam and the outer fortifications, and is under no pressure whatever, but it goes in an orderly fashion into the customary entrances and from there empties into the conduit.

Two aspects of the dam as described by Procopius show the high degree of sophistication in its construction. The upstream curvature or arch form is in common use today, allowing considerable reduction in the volume of concrete dams sited in deep gorges with steep rock abutments, able to take the high lateral thrusts exerted by this type of structure. This seems to have been the case at Daras. The upper sluice gate referred to by Procopius may have been, in effect, a spillway to maintain the top level of the reservoir, allowing excess water to spill over into the riverbed below. The provision of a gate suggests that the reservoir surface could be maintained at two different levels: normally, perhaps, at the lower level, but shutting the gates in flood times to allow the reservoir to store additional water. The lower sluice gate could also have been used as a flood control device, allowing water to be drawn off rapidly to lower the reservoir level in advance of a threatened flood. This may also have been the outlet used to provide water supply to the city. Procopius describes various other hydraulic works commissioned by Justinian, including at Antioch river diversion works and the construction of an immense wall or dam with sluice gates, allowing water to back-up at flood time and to allow the force of the water to be dissipated gradually.

Although remnants of Roman water supply systems survived into the medieval period in Europe, most people drew their water for domestic use from local wells, springs and streams. Wealthier people paid water-carriers to supply them. The common practice of locating natural wells or cisterns to store rainwater in low-lying areas close to cess-pools and latrines, to which water running off the unpaved streets gravitated, carrying refuse and human and animal excrement, led to a rapid spread of epidemics. The communities that often did enjoy efficient water-supply systems were the monasteries, with the Cistercian monks notably active in this work. The monks of the Abbey of Obazine in France diverted water from the River Coyroux into a 1,500 metre long channel, which, for the first 1,250 metres, followed a falling contour of only 0.5 per cent through precipitous country before dropping more rapidly for its final discharge into the monastery. Part of the channel had to be cantilevered out from cliff faces. Canterbury cathedral and its precincts received water conveyed by lead pipes from a source some three-quarters of a mile distant, the water passing through a series of settling tanks before reaching the cathedral. A rare example of a medieval aqueduct with Gothic arches built in 1277 supplied water to Coutance in France. The Black Death which had ravaged Europe in the middle of the thirteenth century hammered home, at least to the more enlightened members of the populace, the need for a clean, unpolluted water supply. Prosperous Flemish towns were among the first to set a good example, while in England the first Urban Sanitary Act passed in 1388 forbade throwing filth and refuse into ditches, rivers and all waters.

When the Normans, in the ninth century, destroyed the Roman aqueduct supplying water to Paris, the citizenry at large made no attempt to restore it, preferring, with the exception of some religious communities, simply to draw water from the Seine or from local wells. Householders who could afford to paid water carriers to fetch water from a well or from the Seine, to keep filled large open vats attached to the houses, supplied from balanced wooden pails on shoulder yokes. The wealthy abbey of Saint-Laurent on the north side of the city, and a considerable distance from the Seine, fed spring waters into a reservoir located in the village of Saint-Gervais some distance away, then ran the water through 125 mm diameter lead pipes to Paris. Another wealthy abbey, also on the north side of Paris, the Saint-Martin-des-Champs, repaired the existing 1.2-km-long masonry Belleville aqueduct to receive water collected from springs. As well as satisfying both abbeys' requirements for domestic purposes and for their extensive gardens, water from the two schemes was also conducted by lead-pipe mains to fountains in Paris for public use. Taking advantage of this, the Convent of the Filles-Dieu sought, and obtained, a concession to pipe water to their building from a public main, to be quickly followed by various nobles, merchants and other privileged persons seeking the same permission. With a little financial or other encouragement in some cases perhaps, the authorities granted many of these requests to allow pipes to be laid to their private properties. As a consequence, fountains often ran dry, forcing local people again to having to either fetch water from the Seine themselves or pay water-carriers to bring it to them. In 1392 Charles VI issued letters-patent annulling these privileges, although allowing his own royal family members to keep them.

By the end of the fifteenth century, sixteen public fountains in Paris received water from the Saint-Gervais and Belleville schemes. Much of the water was distributed by public water-carriers, who had to defer to private inhabitants seeking their own water. Other restrictions imposed on the public carriers meant they could only draw water from the fountain basins when full, and only between sunset and sunrise. To avoid delay they had to keep their yokes over their shoulders while waiting their turn at the fountain. They were not permitted to supply water to certain trades such as dyers and horse-dealers. Heavy penalties applied to anyone washing clothes or other articles in the fountain basin or allowing animals to drink from them. Despite Charles VI's ordinances of 1392, the authorities, perhaps in some cases with financial or other incentives, granted petitions by various nobles, leading merchants and other prominent persons to pipe water direct to their properties from the mains system. By the middle of the sixteenth century this had seriously reduced water flow to the public fountains, leading to the introduction of further regulations requiring that fountains in private premises, with very few exceptions, must be made accessible to the general public or else cut off.

Much of the water supply system in Paris fell into disrepair during the civil war in the last decade of the sixteenth century and, although restored, shortages remained, prompting Henry IV to seek other sources of supply for the abundant water needed for his palaces at the Louvre and Tuilleries. He accepted a timely proposal by a Flemish engineer, Jean Lintlaer, to set up a pump under one of the arches of the just completed Pont Neuf, to raise water from the Seine by means of a wheel turned by the river current and deliver it to the two royal houses. Faced with some opposition by aldermen and merchants to this plan, on the basis of its causing obstructions to the passage of boats, the king reminded them, through his minister the Duc de Sully, that he, not they, had paid for the construction of the bridge. Although the exact features of this installation known as the Samaritaine are lost, it is believed that the famous hydraulic engineer Belidor followed the original design closely when replacing it in 1714. Wooden cribs extended upstream and downstream from the arch, the upstream portion converging to concentrate water flow towards the 5 metre diameter undershot wheel, the shaft of which was supported on the cribs in such a way that it could be moved up and down to accommodate to extreme water levels, such as could occur in flood or drought. Cranks at the ends of the shaft transmitted a reciprocal motion to the ends of 6.5 metre long rocker beams, the other ends of which operated submerged piston pumps. The pumps could deliver about 3 metres cubed per hour with a total lift of 20 metres or so. The king allowed some water, excess to his requirements, to be delivered to a fountain on the island of Cité.

The southern areas of Paris remained poorly catered for and some residents distant from the Seine had to dig wells as deep as 30 metres for water. To remedy this, Henry IV, shortly before his death in 1610, ordered the reconstruction of the old Roman aqueduct at Arcueil, which his widow Marie de Medici eventually saw through to completion in 1623; but the many private permits granted by the authorities meant that the general public benefited very little from the scheme. After much procrastination by the authorities, the 'engineer in ordinary to the King', a man named Joly, was given permission to erect at the Pont Notre-Dame an installation similar to the Samaritaine at the Pont Neuf. After completion, the city purchased the plant and operated it as a municipal undertaking.

Pride of place for the most inefficient works of the seventeenth century go to those at Marly, built in 1682 to supply water for the gardens at Versailles, the grandiose design of which apparently appealed to Louis XIV, the Sun King. Fourteen undershot wheels provided the power to drive 225 pumps in three steps up a 1,200 metre long incline. The system leaked and most of the power disappeared in the clatter of chains, rods and valves running up the hillside. Of more significance than the pumping function was the fact that the Flemish engineer Rennequin Sualem used cast-iron piping, a material which up to then had been used almost exclusively for the construction of cannon. Other similar waterworks in Europe included a system in Augsberg in Germany, which lifted water into towers 40 metres above stream level, and a system built in Toledo in Spain, illustrated by Ramelli, the workings of which mystifies engineers to this day.

A well preserved medieval water supply in Exeter in England derived its water from St Sidwell's Holy Well, situated some distance outside the east gate of the city and owned by the Dean and Chapter of the cathedral. In 1266, St Nicholas's Priory obtained a grant entitling them to one third of the water from the well, with the city and cathedral dividing the remaining share equally. Tunnels excavated into the underlying rock to conduct the water into the city had rather haphazard cross-sections, varying in height from 1 metre to 4.5 metres and in width from 0.5 metres to 1 metre. Mostly rounded bottomed, many were also lined with rough barrel vaulted stonework. Water was conducted to a central point and thence conveyed by three channels to the cathedral, the city and St Nicholas's Priory – the last two paying the cathedral authorities 8*s* a year for their supply. In the fifteenth century the city, requiring larger amounts of water, laid lead pipes about 50 mm in bore to access another source in the St Sidwell's area.

In the west country of England a number of medieval towns situated below rising ground tapped suitable sources of water at a higher level such as springs, streams or wells and conducted it into the town by means of a leat or open channel following a gently falling gradient. Plymouth was a notable example. In the reign of Elizabeth I, its strategic importance as a port required that it should cease to rely solely on wells for its water supply, and by virtue of an Act passed in 1585 received permission to bring water from Dartmoor to the town by means of a leat for the primary purposes of watering ships, protecting against fire and scouring the harbour, silted up by tinworks. No less a figure than Sir Francis Drake promoted the scheme to throw a weir across the River Meavy and divert water into a leat, the Corporation being authorised by the Act to

> digge and myne a diche or trenche conteynenge in breadth between six or seven foote overall the lands lyeing betweene the saide town of Plymouth and anye parte of the saide river of Mewe als Mevye and to digge, myne, break, bancke and cast upp all and all mener of rockes, stones, gravel, sande and all other lets in any places or groundes for the convenient or necessaire conveyenge of the same river to the saide town.

Drake, now MP for Bossiney in Cornwall, and who had been mayor of Plymouth in 1581, not only promoted the scheme but also served on the Select Committee that considered and approved the scheme. For a year or two after approval of the Bill, Drake answered the call of his country, 1587 seeing him in Cadiz irritating the King of Spain and 1588 finding him in the English Channel sorting out the problem with the Armada.

With these small matters behind him, he could now turn his attention once again to Plymouth's water supply. He contracted with the town to execute the works himself for a sum of £200, with an additional sum of £100 for compensating landowners through whose properties the leat passed. The corporation covered its own costs and the stipend of one Robert Lampen, appointed as engineer for the project. His contract also gave him the right to establish, and operate for sixty-seven years, six mills along the course of the channel, a concession bitterly opposed by mill owners on the Meavy, who took their case to Parliament. A Select Committee of the House (chaired by Sir Francis Drake!) rejected their appeal.

Drake himself cut the first sod in 1589 and, on 24 April 1591, to mark the completion of the project, reputedly rode ahead of the first flow of water along the channel on a fine white horse and after the arrival of the water into Plymouth dipped his scarlet cloak into

A sixteenth-century conduit in Westcheap, London, showing the variety of vessels used to convey the water to private premises, the larger tankards used by watermen and smaller ones commonly by house servants.

the channel amid much celebrating, firing of guns and feasting. The need to follow a gently falling contour required a leat over 27 km in length, although the direct distance from source to town measured only some 19 km. The sides were lined with granite to protect it from scouring and in later times the bottom was concreted. Wooden and lead pipes distributed the water in the town, with the general public allowed to take water free of charge from the conduits or conduit houses in various parts of the town. The public conduits fed water to wooden troughs, later replaced by stone. For 4s per annum, wealthier inhabitants could have their own private supplies. The scheme served with little modification as Plymouth's water supply for 300 years. As part of his contract, Drake secured a 67-year lease on six water mills along its course.

Stow, in his *Survey of London*, claims the capital to have special wells 'sweet, wholesome and clear', the most famous being Holy well, Clarke's well (Clarkenwell, now Clerkenwell) and Clement's well. Holy well he states to be much decayed and marred with filthiness, which rather contradicts his previous observation, but Clement's he claims to be clean and always full. He describes Clarkenwell as 'curbed about square with large stone', and to have been named after the parish clerks in London, 'who of old times were accustomed there yearly to assemble, and to play some large history of Holy Scripture'. It appears that nobles and even the King and Queen on occasion attended to witness these plays at Clarke's or nearby wells.

1582 proved to be a momentous year in London, when Peter Morice (or Morris), who had worked with pumping machinery in Germany, completed the construction of an undershot waterwheel to utilise the tides sweeping through the arched openings under London Bridge. Morice had proposed his project as early as 1574, but the City fathers, with no experience of any such machinery, procrastinated until Sir Christopher Hatton, a favourite of Queen Elizabeth and later Lord Chancellor of England, who had previously used Morice's services, exercised his influence to obtain permission for Morice to install his equipment under the most northerly arch. A sketch by John Bate made in 1635 showed the end of the rotating shaft of the undershot wheel, projecting through its supporting timber frame, to have fixed to it a crank driving a long connecting rod giving semi-rotary motion to a vertical disc, which, in turn, gave a rocking motion to a half-disc above it by means of driving chains linking the two. The rocking motion of the half-disc imparted rectilinear motion to vertical piston rods for two pumps. This remarkable arrangement meant that the pumps operated regardless of which direction the wheel was turning and consequently utilised both incoming and outgoing tides. In order to demonstrate the effectiveness of his pumps, Peter Morice threw a jet of water over the steeple of St Magnus Church, which impressed the municipal authorities sufficiently to grant him and his heirs a 500-year lease of the two northern arches of London Bridge at an annual rent of 10s. Morice built his second machine in 1583. Not surprisingly, the Brotherhood of Water Bearers did not view Peter Morice's pumps with the same enthusiasm as the municipal authorities, despite assurance by the Lord Mayor (surely tongue in cheek) that they would have more work in future, not less. The pumps raised the water through a 300 mm diameter mains pipe to a tower top reservoir nearly 40 metres high, as described by John Bate, 'these pipes carry the water to the top of a Turret neare adjoining unto the Engine and there being strained through a close wyer grate it decendeth into the maine wooden pipe which is layd along the street and into it are grafted divers small pipes of led [lead] serving each of them to the use and service of particular persons'. Morice's wheels rotated six times a minute, their noisy rattling, creaking and groaning apparently insignificant compared to the female shrew in a contemporary play:

> Oh, terribly
> Extremely fearfully! The noise of London Bridge
> Is nothing near her.

Some of the nearby inhabitants might have thought differently!

In 1701 Morice's descendants sold their rights to Richard Soames, who employed hydraulic engineer George Sorocold to reconstruct the machinery rebuilt after the Great Fire of London in 1666, which had fallen into a state of bad repair. Sorocold had already constructed similar works in several other towns in England, such as Derby 1692 and Bristol in 1695, and a wheel and pumps he installed at Bridgenorth in 1706 still operated well into the second half of the nineteenth century. Ultimately, waterwheels driving pumps occupied the five most northerly arches of the bridge, the last being installed by Smeaton in 1767. Demolition of the bridge in 1831 saw the end of this source of water supply in London, but an Act of Parliament passed in 1822 allowed the proprietors to transfer their rights to the New River Company for an annual payment of £3,750 for the remainder of Morice's 500-year concession.

The London Bridge works served only a limited area of a metropolis already growing beyond the city boundaries, joining up with Westminster and expanding into the adjacent countryside. In 1604 James I granted Edmund Colthurst of Bath a patent to dig a river

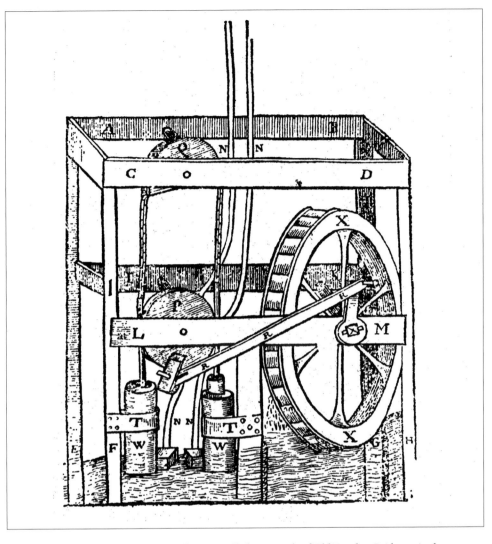

Peter Morice's waterwheel-driven machine, installed in an arch of Old London Bridge, raised water to a tower-top reservoir 40 metres high. It pumped water during both incoming and outgoing tides.

not more than 6 feet wide to bring water to London from springs in Hertfordshire and Middlesex, the work to be completed within seven years. For this permission, Colthurst was to pay the King £20 per annum and the water was to be distributed one-third to houses in London and Westminster and two-thirds to cleansing various city ditches. By 1605 he already found himself in financial difficulties, having excavated about three miles of channel, and the work ceased as the City authorities declined to make any financial contribution. In October 1608, having taken on a number of partners, Colthurst put forward alternative proposals to do the work, either with the city contributing £2,400 in exchange for two-thirds of the water or contributing nothing with Colthurst and his partners taking all the profits. A committee appointed by the City Council appeared favourably disposed to the second proposal, but in April 1609 the City authorities came to an agreement with Hugh Myddelton (later to become Sir Hugh Myddelton), a wealthy goldsmith and Member of Parliament, to perform the work. In his latter role he had sat on an advisory committee in 1606 considering a Parliamentary Bill which, among other things laid down a procedure for settling matters of compensation to landowners and allowed the width of cut to be increased to 10 feet.

Myddelton, his brother Robert, and other similar worthies probably constituted the partners enlisted by Colthurst and, realising that the authorities would only approve the work if headed by a man of such stature, he probably accepted this arrangement as inevitable. Myddelton appointed Colthurst as engineer for the work and assigned to him for his lifetime four of the thirty-six Adventurers' shares. The scheme was based on drawing water, as Colthurst had intended, from a natural spring at Chadwell in the Lea valley, a short distance above Ware in Hertfordshire, the flow to be supplemented from a spring at Amwell some two miles downstream. The channel, 10 feet wide and with an average depth of 4 feet, was excavated to follow a dropping contour with a gradient of only 2 inches to a mile, thus requiring a total length of 38¾ miles from its source to a circular pond or cistern into which it fed the water at Islington, north of London, almost twice the direct distance of 20 miles between the two. Additional water was fed into the channel by pumping from wells along its length. Embankments up to 8 feet high carried the channel over low lying ground and, where necessary, roads or streams were taken under the channel and the channel carried across these on arched aqueducts or in timber troughs lined with lead. One of these troughs, supported on timber arches fixed in the ground (later replaced by an embankment), had a length of 660 feet and a depth of 5 feet and another, 460 feet long and 17 feet high, carried the channel over a valley near where it entered the parish of Islington. Other constructions included two short brick tunnels and more than 150 timber bridges to maintain existing field and other accesses.

The work had hardly begun when owners and occupiers of the land through which the planned route of the New River passed suddenly found their voice and presented a Bill to Parliament to stop the scheme, alleging that their meadows would be turned into bogs and quagmires, to the utter ruin of many poor men, and the highway from Ware to London would be rendered impassable. The appointment of a committee to report on these objections came to nothing, because Parliament did not meet for four years. The channel passed through the Royal Park at Theobalds, near Enfield, but unlike some other landowners James I strongly supported the scheme, taking a personal interest, and even provided much needed financial assistance for a share of the profits. His enthusiasm for it may have been a little dampened (literally) sometime after its completion. Out riding in the park at Theobalds after dinner, in winter, his horse stumbled and threw him into the slightly frozen river and he disappeared under the water with only his boots showing. His accompanying staff quickly hauled him out and despite expelling much water 'from his mouth and body', he remounted his horse and returned to the royal residence, where he quickly recovered.

Irrigation

When Islamic forces defeated the Sassanians in 636 and again in 641, they rejected the powerful Sassanian beliefs in the competing sects of Zaroastrianism and Manichianism, but nevertheless absorbed Sassanian social and political systems as well as their technology. This gave them the foundations upon which they built and innovated to become outstanding technologists in many fields, including hydraulic engineering. Irrigation in Sassanian times reached its peak in the sixth century AD, as evidenced by the Nahrwan Canal, 450 km long, taking off from the left-hand bank of the Tigris some 200 km above Baghdad and rejoining the Tigris another 200 km below Baghdad. Part of a complex irrigation scheme, it captured left-hand tributaries flowing into the Tigris, the complete system requiring various dams and control works. It is likely that parts of the Nahrwan scheme were constructed or reconstructed during the period of the Abbasid Caliphs who ruled the Islamic world from the glittering city of Baghdad between 762 and 1258. Dams were built across the Tigris and Adheim rivers to the north of Baghdad, mostly for the purpose of maintaining or improving the Nahrwan scheme, the most impressive dam being that on the Adheim River to divert the river water into the Nahr Batt canal, which fed into the Nahrwan Canal. The 175 metre long dam was constructed of cut masonry blocks and had a maximum height of about 15 metres with a trapezoidal section 3 metres wide and the crest and 15 metres wide at the base. Water-tightness was achieved by pouring molten lead into matching grooves in the faces of adjoining blocks.

In addition to extending the Nahrwan scheme, the Muslims constructed several canals to the south of Baghdad connecting the Euphrates to the Tigris at a slightly lower level, the lands irrigated by these supporting a thriving agriculture. The largest and most important of these, the Nahr Isa, close to Baghdad itself and able to carry shipping, fed a number of parallel secondary channels and these in turn supplied many small transverse cuts. A system of dams on the Euphrates forming the intake for the canal overtopped on at least two occasions, causing serious flooding in Baghdad. The town of Basra on the Shat-al-Arab, set up as a military encampment in 638, rapidly grew into a large city and a centre of commerce and overseas trade. It spanned a network of irrigation canals and became a centre for the agricultural industry.

As Islam spread into Asia and across North Africa the Muslims took their knowledge of irrigation methods and machinery with them, although in some places, notably Egypt, they took over existing, well-developed systems. On the Nile they erected a Nilometer, still in existence, consisting of a stone column marked off in cubits, the rise in water-level determining the land tax to be paid to the Sultan. They grew new crops. Rice was introduced into Sicily and Spain, whence it made its way northward to be grown on the plains of Pisa and Lombardy by the fifteenth century. Oranges, cotton and sugar cane were among other new crops grown by the Arabs, but they did not confine themselves to the use of irrigation just to grow foodstuffs, as witness the Generaliffe gardens high on a hill in the cooler airs overlooking the Alhambra in Granada. Laid out in 1230, it comprised an integral part of the Summer Palace of Sultan Mohammed ben Alhamar. Its subtlety of detail and its beautiful site makes it one of the world's most beautiful gardens

The Generaliffe gardens in Alhambra were first laid out in 1230 as part of the Summer Palace of Sultan Mohammed ben Alhamer.

and a leading tourist attraction today. The Islamic engineers used water with great finesse to create dancing finger jets, surging fountains, gurgling streams and strips of water lending perspective and enchantment to the gardens.

The Alcazar in Seville also boasts a magnificent garden of typical Moorish design, although in fact it was the Christian King Pedro the Cruel who commissioned the Arab gardeners to lay out the garden on a citadel site, exploiting the River Guadalquivir to provide water to irrigate it. Shade from burning sun, fragrance and tranquillity, those unmistakeable hallmarks of Moorish gardens, come together in perfect partnership here, the cool shadows of tree-lined alleys contrasting with the intermittent patches of sunlight and sparkling waters of the irrigation channels. The gardens are a riot of vegetation set out in separated, but connected, geometric patterns serving as an example for the formally laid out gardens in northern Europe in the seventeenth and eighteenth centuries. Here in Seville a bewildering array of plants includes clipped box hedges, stands of cypresses, myrtles and oleanders, jasmines and orange blossoms spicing the air with their unique fragrances, climbing roses bedecking the date palms, blue and purple irises, plantains and daturas with their large perfumed bells.

In general, the Moslems were as well aware as the Romans of the need for a good water supply and they restored some Roman systems that had fallen into decay. Unlike the Romans, however, the systems they built often combined the supplying of water both for irrigation and urban usage. Waterwheels found widespread use in Islamic countries for lifting water to a level where it could flow by gravity to the fields for irrigation or into cities, such as Córdoba, which was supplied by a noria (waterwheel) from the Guadalquivir River. The river both supplied water and turned the wheel. Diversion dams up to 10 metres or so in height were also used to raise water and feed it into channels supplying irrigation or urban needs. These dams commonly consisted of a core of masonry rubble set in lime mortar faced upstream and downstream with cut and fitted masonry blocks. The solid construction allowed flood waters to pass over the tops of these structures.

Irrigation had been practised in Italy long before medieval times, initially by the Etruscans, who handed over their knowledge to the Romans. Much of the land put under irrigation by the Etruscans had been reclaimed by drainage, and they knew very well how to collect, store and channel water to grow grain and fruit crops and to nurture their grapes. Grape seeds found in Etruscan graves in the Chianti district indicate that they probably introduced wine growing into Italy.

The name Naviglio Grande suggests a transportation canal; but in fact this 50-km-long canal, which can be regarded as a forerunner of modern artificial waterways in Europe, originally served to provide irrigation water to farmers along its length, and to supply the moat surrounding the medieval town of Milan. Irrigation was already being practised by Cistercian monks in the areas around Milan in the twelfth century, obtaining water by diverting small streams and by constructing connecting channels. Originally known as the Ticinello, the Naviglio Grande was constructed between 1179 and 1209, taking its water from an intake on the River Ticino, near Casa della Camera, running initially in a south-easterly direction almost paralleling the Ticino to Abbiate, then east to Milan, experiencing a drop of 33.5 metres along its length. Unfortunately, no contemporary written records of the work survive. Giovanna Battista Settala, writing four hundred years later, states that it had a width of 70 braccia (42 metres) at its mouth, diminishing to 25 braccia (15 metres) when it reached Milan. At its source, blocks of stone laid dry to a height of 2 metres protected the right-hand bank, followed by a pile palisade to give protection for the next 3 km. Elsewhere the banks, probably some 2 metres high and 10 metres wide, consisted of the spoil taken from the excavation and set back to provide a good berm to protect the stability of the banks and increase the flow capacity of the canal. Canal slopes varying from almost nothing to a maximum of only 1:1,100 ensured low flow velocities. Six water reliefs at suitable locations drained excess water from the canal at times of flood into intersecting streams. Weirs placed in the channel raised the water level to enable irrigation water to be extracted and in some cases also served to drive waterwheels.

Irrigation had become established practice in the south of France in medieval times, to the extent of a proposal being made in the twelfth century for a canal drawing water from the River Durance to irrigate the fertile plains of Provence around Arles. For some four hundred years it remained a dream. In the sixteenth century, Adam de Craponne, an engineer and descendant of a Pisan family that had settled in the area, undertook the task, having been given permission by the king, Henry II, to divert water from the Durance into the plain of Crau. Having completed his surveys in 1551, his attentions became temporarily diverted by the king requiring his services as a military engineer, but with these duties performed he returned to complete his canal, actually little more than a ditch, to the town of Salon. His careful alignment of the canal around the contours to keep water velocity below that which would erode the banks unfortunately resulted in a totally inadequate flow. Although he bore most of the project costs himself, he still experienced threats and abuse from potential beneficiaries; but these vilifications turned quickly enough to rejoicing when a wealthy landowner of Provence advanced the money for him to enlarge the canal, bringing a copious quantity of water to Salon, just in time to save the local crops from severe drought. Extension of the canal to Arles had to wait until after Craponne's death or, more specifically, his probable murder in 1575. Henry III had sent him to Nantes to report on foundations recently laid for new fortifications and, anticipating his probable condemnation of the work, the contractors first tried to win his favour by flattery and the offer of presents, which he rejected, so there is good reason to believe that they poisoned him.

Power from Water

As with many other areas of technology, the Moslems absorbed the methods of power generation developed by the ancient and classical world, and through them this knowledge eventually spread throughout Europe – partly by direct contact, particularly through Spain and Sicily, and partly conveyed back to their homelands by returning Crusaders. All basic types of waterwheel found use in medieval Islam: the undershot wheel, the overshot wheel, breast wheel and the horizontal wheel. Medieval Islamic waterwheels, usually undershot, served mostly for milling grain, but exceptions included mills to drive trip hammers for paper making. In 751 AD Chinese paper makers, made prisoners of war by the Arabs at the battle of Atlakh, set up paper mills in Samarkand making paper from flax, linen or hemp rags. The technique spread rapidly through the Islamic world. Other known uses included processing sugar cane and sawing timber.

Medieval Europe saw a rapid rise in population and an explosion of water mills, not only in numbers but in their application to a wide range of industrial uses. In addition to the milling of grain, these included paper manufacture, iron forging, ore crushing, sawmills, tanning and the fulling of cloth, and gunpowder production. Water power was even adapted to such tasks as sawing off timber piles underwater for bridge foundations, as illustrated by Villard de Honnecourt. This use of waterpower initiated the first widespread mechanical means of power production and the Cistercian monks figured large among the major industrialists exploiting it for many industrial purposes. Much of the work associated with these undertakings fell to lay brothers admitted to the monasteries on a footing equal to that of the monks, but with no spiritual duties or obligations and thus free to pursue commercial enterprises. Their monasteries operated hundreds of mills throughout Europe, collecting rich pickings from their tenants, who were obliged to bring their corn for grinding, and cloth for fulling, to the mills. Manorial landlords made similar demands on their tenants, many of whom had to trudge miles to the mills and then pay high rents when in fact they could readily have done the grinding and fulling in their own homes. Some of them did. Fighting followed. Houses were searched and millstones seized. The bitterness engendered contributed much to the peasant uprisings of the fourteenth and fifteenth centuries.

Many mill owners in Europe favoured inefficient, but cheap to construct, undershot waterwheels, which, unlike overshot wheels, did not need the construction of special water races to bring the water to the mill at an elevated level. The undershot wheels could be located directly in the stream or in a localised channel diverted from the stream. Where a mill could not be conveniently located on the bank of a river, floating mills offered a solution which had been in common usage since Roman times. The restricted span openings of typical medieval masonry bridges presented a temptation difficult to resist and they often hosted waterwheels, making use of the fast flow of water through them. Scour problems, already bad enough with wide, closely spaced piers, were greatly exacerbated, leading to weakening or undermining of pier foundations.

Waterpower could also be generated by the construction of a dam or weir across a stream to raise the level of the water, which could then be released to flow rapidly along an open channel to the mill to drive an undershot waterwheel. An example of a masonry dam some 2.5 metres high built by the Moslems in Spain to provide waterpower can still be seen across the Guadalquivir at Córdoba. Constructed with a zig zag alignment across the 300 metre wide river, probably to increase its flow capacity, it had a mill house located at each of the three downstream points of the zig zag. The raised water also drove a huge noria or waterwheel built into the bank of the river to lift river water to the level of an elevated aqueduct that conveyed water into the city.

Tidal fluctuations provided another source of medieval waterpower through the construction, in tidal creeks or rivers, of structures incorporating swinging gates that allowed water to flow into the creek, but closed under the pressure of the ponded water when the tide dropped. The raised water then flowed through millraces to drive the waterwheels. It says much for the judgement of the mill builders of the time in siting their structures that they located a number of tidal mills in the estuary of the River Rance in France, where the world's largest modern tidal-power scheme now operates.

France, too, can lay claim to the world's first public limited company, the assets of which comprised a series of three dams and associated watermills on the River Garonne, built in the second half of the thirteenth century. Each dam consisted of two parallel rows of oak piles driven into the river bed, with the space between them filled with timber, clay and stones. Breakwaters protected the dams from floating debris. As the raising of the water level in the lowest and middle dams interfered with the operation of the middle and highest dams, respectively, innumerable lawsuits resulted. Eventually, the owners of the lowest scheme, Le Bazacle, contrived to take over the middle scheme, La Paurade, and by 1374 Le Bazacle and the highest scheme, the Chateau-Narbonnais, had formed into two companies owned by wealthy Toulouse shareholders. Agreements drawn up between the companies served for nearly eight hundred years before the French government absorbed the companies into its nationalisation programme after the 1939–45 war.

The Domesday Book, compiled by William the Conqueror's Commissioners in 1086, lists an incredible 5,624 watermills in 3,000 communities in England south of the Trent and the Severn. They became an integral part of everybody's life and featured in the writings of the day. By the end of the fourteenth century most major towns and cities in Europe had substantial numbers of water mills; Paris, for example had at least fifty-five exploiting the waters of the Seine. Practically every sizable village boasted at least one water mill. In *The Reeve's Tale*, written at this time, Chaucer dwells, albeit lightheartedly, on the dishonesty of a miller:

> At Trumpyngtoun, nat fer fro Cantebrigge,
> Ther gooth a brook, and over that a brigge,
> Upon the whiche brook ther stant a melle;

Or, perhaps, in modern verse:

> At Trumpington, not far from Cambridge,
> There is a stream, and over it a bridge,
> And upon the stream there stands a mill;

There then follows descriptions of the bullying miller, his upper-class wife (brought up in a nunnery), their 20-year-old daughter and their 6-month-old baby. The problem arises because the miller had sole rights to grind wheat and malt in the vicinity, which included a large Cambridge college known as Solar Hall (apparently a nickname for King's Hall):

> Ther was hir whete and eek hir malt ygrounde.
> And on a day it happed, in a stounde,
> Sik lay the maunciple on a maladye;
> Men wenden wisly that he sholde dye.

For which this Millere stal both mele and corn
An hundred tyme moore than biforn;
For therbiforn he stal but curteisly,
But now he was a theef outrageously,

Or, perhaps, in modern verse:

They [the College] had their wheat and malt ground at this mill,
But on one day it happened, as it will,
The College bursar lay sick in bed;
Some thinking him quite close to dead,
The miller now stole flour and corn from the College store
A hundredfold more than he had before;
For then he stole respectfully,
But now he thieved outrageously.

In continuing his tale, the Reeve relates how two young scholars then persuaded the College Warden that they should take the next bag of corn to the mill and oversee its grinding; but with their attention thus occupied, the miller released their horse, who galloped off to the nearby fen, where there were wild mares. They gave chase and did not return to the mill until late at night, giving the miller the opportunity not only to steal half a bushel of their flour, but to have his wife bake it into a cake. Too late to return to the College, they dined with the miller and his family, with much drinking of ale all round, and retired to bed, all sleeping in the one and only large bedroom. With the miller 'snoring like a horse', one student quietly made his way in the dark to the daughter's bed and the other, by subterfuge, got the miller's wife, unknown to her, into his bed. All but the miller had a joyful night. When one of the scholars inadvertently related to the miller in the morning how much he had enjoyed his daughter, the miller attacked him, and the wife entered the fray and brained her husband with a stick when aiming for the scholar. The students immediately took their horse, flour and the cake and departed.

By 1700 at least half a million waterwheels throughout Europe provided power for a multitude of tasks. Furthermore, European colonists had set up waterwheels overseas for a variety of purposes, such as crushing sugar cane in the West Indies. Widespread use of the waterwheel continued well into the second half of the nineteenth century, despite the advantages of the steam engine with its compactness and mobility and freedom from the vicissitudes of nature, in the form of droughts, floods and ice, which could halt the supply of water power and even badly damage the units. The development of the efficient all-iron water wheel helped to prolong its life and highly efficient water turbines are an important source of power today, their immobility no longer a problem as the electricity they generate can be conveyed to its point of usage, wherever that may be. Environmental concerns are now leading to the investigation of other possible ways of harnessing water power, such as exploiting wave action or underwater tidal currents.

Land Drainage and Reclamation.

The principles and practices of land reclamation by drainage or exclusion of water were well established by the middle ages, with a number of ancient civilisations having implemented substantial schemes. As early as the seventh century BC the area that became known as the Forum in Rome was systematically drained by open trenching, the main trench being subsequently converted into Rome's famous Great Drain, the Cloaca Maxima. At around the same time the Etruscans drained parts of the Campania, south of Rome, by means of trenches about 1.5 metres deep cut into the volcanic soil, to create rich agricultural land. In 325 BC the Greek engineer Crates put in hand a scheme to drain Lake Copais, a vast reed swamp some 65 km north of Athens, which included the construction of a tunnel over a mile long. He had the work well in hand when Alexander's military activities brought it to a halt. The work finally resumed in modern times and with its eventual completion in 1890, the former lakebed is now farmland.

In Roman times, according to Pliny, the inhabitants of the Dutch coastal provinces lived on patches or mounds of high ground standing above high tide level. To some extent, these appear to have been built up by the deposition of refuse, but after about AD 500 the sea level rose by 100 mm or so per century in relation to the land, which exceeded the rate of build-up of the mounds, thus making necessary the construction of perimeter water defences that marked the beginning of Dutch dyke construction. With the passing of time, dyke construction extended to outer coastal defences and the enclosure of low-lying land from which water was released at low ebb tide. Automatic sluices evolved, activated by water pressure, closing under a rising tide level and opening to allow water to flow out under a falling tide. These drained enclosures became the first polders. The outer coastal defence works consisted largely of reducing the length of coastline, and hence the lengths required to be defended, by closing the bays, creeks and gaps. By the end of the fourteenth century most of southern Holland enjoyed coastal protection and the three centuries between 1200 and 1500 saw the creation of over 100,000 hectares of new agricultural land.

Land reclamation achieved by embanking the perimeter of an area and releasing water through sluices had the drawback that it could not be used to drain land lying below ebb tide level. Reclamation of such land required pumping, and wind power provided the answer. Scoop wheels and Archimedean screws driven by windmills lifted the water from low-lying meres over the enclosing embankments (ringdijks), depositing it into ditches (ringvaarts) which carried the water into rivers, canals or into the sea.

Subsidence of land relative to the sea occurred progressively through the medieval period, forming the Zuider Zee, a shallow inlet of the North Sea, which reached its greatest extent around 1300. These relative changes in level also constantly threatened to undo human efforts in establishing sea defences, over 50,000 people drowning in floodwaters during a storm on 14 December 1287. The inhabitants could never relax their vigil against the aggressive and unpredictable sea. Any deterioration of the dykes could lead to disastrous consequences, as witness the St Elizabeth's Flood in 1421, when waters driven by a November gale opened two large breaches in dykes at Wieldrecht which had been allowed to fall into a poor state of repair. The waters of the River Maas poured through, engulfing several villages and drowning thousands of people and many cattle. The All Saints' Flood of 1570 effected a considerable change to the coastline.

The Dutch developed an expertise in the use of locally available materials for the construction of dykes, relying on clay to form the core or body of the dyke with various means for protecting the slopes. Glacially deposited clay proved most suitable for core construction where available, but poorer quality clays often had to suffice and these required greater slope protection. Where the core consisted of good quality clay and the slope to be protected did not directly face the open sea, protection sometimes consisted simply of a slope finished to one vertical to 1.5 horizontal planted with grass. For conditions demanding greater protection, a common method consisted of placing bundles or bales of straw, osiers, seaweed or reeds against the face of the dyke; by the fifteenth century more substantial methods had been developed, based on either a single row of driven timber piles or a double row with the space between the rows infilled with stones or faggots.

The sixteenth and seventeenth centuries saw a change in emphasis between land reclaimed by dyking against the sea and the polders created from the drainage in inland meres. Between 1540 and 1565 the former accounted for only 1,350 hectares out of a total of 3,560 hectares reclaimed, but between 1615 and 1640 it accounted for 19,000 hectares out of a total of 25,500 hectares. One of the leading figures in the draining of the meres was Jan Adriaanszoon Leeghwater (1575–1640), the son of a carpenter who, although not introducing any new fundamental ideas into windmill design, constructed them in such a way that they operated more efficiently and pumped greater quantities of water than previously. He also developed to its greatest extent the system of multi-stage water lifting in which two, three or four scoop wheels worked in series to lift water from the lowest level of the polder to the encircling drainage ditches. Leeghwater used this system to drain the Beemster mere between 1608 and 1612, a work commissioned by the Dutch East India Company and which established his reputation. He participated in most of the major reclamations undertaken during his working lifetime, including the Purmer (1612), the Heer Hugowaard, the Widje Wormer (1626) and the Schermer; but he never achieved his greatest ambition to drain the Harlemmermeer, a 16,500 hectare area of lakes and swamps, the waters of which even threatened Amsterdam and Leiden during periods of storm. He estimated that 160 windmills would be required to drain it. In fact, the project remained in abeyance until 1840, when drainage started by order of King William I, using not windmills but steam driven pumps, which in 12 years dried the bed of the mere now occupied by Schipol (meaning a shelter for ships) Airport, the aircraft landing on runways 4 metres below sea level.

The reclamation of land recovered from the sea presented problems even after completion of drainage, not the least of these being the removal of salt from the saturated soil. This was effected simply by the action of rainwater passing down through the soil, carrying the salt-laden water with it into deeply incised trenches from which the water was pumped out of the polders. The annual precipitation in the Netherlands is not particularly high, some 710 mm, and in summer evaporation exceeds rainfall, so the salt is drawn up by capillary action, replenishing in some measure salt leached out during the winter. Consequently, it took several years to remove the salt, although some crops such as grains and grasses could be planted earlier than salt-sensitive legumes and potatoes. Health also posed a problem in early reclamation schemes. Where soil did not drain freely and quickly, pools of water formed and remained standing, thus becoming ideal breeding grounds for the malarial mosquito. Severe malaria epidemics occurred among the pioneers of Dutch reclamation projects.

The Church commissioned much of the early reclamation work carried out in the Netherlands, the Count of Flanders having granted to the abbeys and chapters in his maritime region new lands still exposed to the sea. Religious houses in other parts of Europe took up the challenge, often employing the skills of the Netherlanders. Archbishop Frederick of Bremen called in their expertise to reclaim swamplands north-east of his see in 1106. The Cistercians were active in drainage and irrigation, particularly in Germany, and the Benedictines in France. Later, the initiative for new reclamation works passed to the land-owning nobles or kings, town councils and large private concerns, notably the Dutch East India Company. It was across the North Sea in East Anglia, however, that Dutch expertise had its greatest impact through the efforts of Cornelius Vermuyden (1590–1677), who may well have worked with Leeghwater as a young man and who would certainly have been influenced by the treatise of another great Dutch 'dyke-master', Andries Vierlingh, written around the middle of the sixteenth century.

The fenland of eastern England, a rich market gardening and farming area, embraces some 1,306 square miles or 260,000 hectares to the north of Cambridge, mainly in Cambridgeshire, Norfolk and Lincolnshire. Its flat landscape, which bespeaks of its very recent existence in geological terms, as a low-lying swampy area, is relieved by islands composed of hummocks of the harder underlying Jurassic and Cretaceous clays and sands protruding up through the softer recent (10,000 years or so) deposits of soft clays, silts and peats. Although now surrounded by farmland, these hummocks were once genuine islands in a landscape of water and reeds. Major towns like Ely, March and Soham occupy such islands.

The uplands bordering the fens comprise chalk to the east and south-east, and the older Jurassic and Cretaceous deposits to the south-west and west, which in turn are bounded on the west by much harder limestones. The topography owes much to the glaciations that extended into southern England, depositing a carpet of unstratified boulder clay in some areas, but the huge quantities of water released by the melting ice gouged out valleys and basins, such as the basin occupied by the fenland, formed from the erosion of the Jurassic and Cretaceous clays and sands, leaving relatively untouched the harder chalk to the east and limestone to the west. To the north the fens are bounded partly by the Wash, an inlet of the North Sea. At one time the chalk uplands of Lincolnshire extended across the Wash to link with the chalk lands of Norfolk, but the North Sea broke through this barrier, allowing the seawater to invade and erode the fen basin.

Deposition of the alluvial sediments now filling the fen basin commenced subsequent to the end of the last glaciation some 10,000 years ago. Silts deposited by the sea along the seaward margins gradually built up and barred ready access to the sea for the waters pouring into the basin from the 6,000 square miles of bordering uplands. Swamp vegetation proliferated in the non-saline waters and gave rise to thick peat beds, through which rivers pursued meandering paths. As a result, the fens can be broadly divided into the silt fens occupying the seaward area and the peat or black fens occupying the landward areas, although this is something of a simplification because the relative extent of seawater and fresh water has fluctuated, giving intercalations of soft clays, silts and peats.

During the Roman occupation the fenlands, particularly the silt fens, supported a substantial population engaged mostly in growing corn and raising cattle and sheep, which suggests that they introduced effective drainage measures, although there is evidence that the fens enjoyed a relatively dry period at this time. A drop in land level in late Roman times, with consequent flooding of larger areas, may have contributed to a lack of any sustained attempts at drainage between about AD 450 and AD 1300, despite some recovery in land levels during this time. Attempts at localised drainage were made by some of the many monastic establishments set up between the seventh and eleventh centuries, but these settlements suffered grievously at the hands of the Viking and Danish predators.

A plethora of truths, half-truths and legends abound about the Fens at these times, which is hardly surprising in view of their remoteness and inaccessibility. A parting of the reeds reveals such diverse characters as the aesthetic monk St Guthlac, who feared his fellow men much more than the 'strange and uncouth monsters' believed to inhabit the Fens; Tom Hickathrift, a 7-foot giant, supposedly freed the country from savage beasts and outlaws; Hereward the Wake may have existed and, with his small band of followers, defied William the Conqueror from a hideout on the Isle of Ely, until betrayed by the Ely monks; and King John in 1215, while hastening northwards from Kings Lynn to repel a Scottish army, attempted to cross a tidal estuary just north of Wisbech and in the process lost a wagonload of treasure which sank into the soft mud, where it has never been found to this day, even though the estuary is now dry land. Perhaps it is awaiting discovery by some enterprising amateur treasure seeker.

A drop in land levels relative to sea level in the thirteenth and fourteenth centuries led to severe flooding in the Fens and the loss of human life as well as large numbers of cattle and sheep. Wisbech was largely destroyed in 1260. In 1268 the central government appointed Commissions of Sewers to enforce the upkeep of dykes and drains and to impose fines on farmers who allowed their cattle to damage the banks. In 1490 John Morton, Bishop of Ely, cut off a length of the River Nene by means of a straight cut from Stanground, near Peterborough, to Guyhirne, where the water rejoined the Nene to flow through Wisbech to the sea. He thus invoked the important principle of shortening a watercourse to increase the water velocity and improve its scouring action, which later became the basis of much more extensive work by Vermuyden.

Morton earned a place in history when, as a Privy Councillor, he argued that merchants displaying outward wealth should contribute to the Exchequer, because they clearly had money to spare, while those that lived in a miserly manner obviously saved much money and were thus well able to contribute as well. This became known as Morton's Fork.

During the Pre-Reformation, the religious houses exercised a dominant influence over the administration of the Fenland and all its works, including the rather limited amount of drainage carried out during this period. Following the Dissolution of the Monasteries in 1539, many of the large tracts on monastic lands were broken up and sold, thus increasing the number of land owners and in consequence diluting the responsibilities for the maintenance of banks and channels. Throughout the remainder of the sixteenth century and the early seventeenth century flooding became more frequent and more damaging, but the magnitude of the problem, together with the hostility of the local people deriving their living from the fish and fowl inhabiting the meres, thwarted any serious attempts at drainage until 1630, when King Charles I asked Francis, 4th Earl of Bedford, a large land owner in the area, to undertake the drainage of the Fens. Under the terms of the Lynn Law Agreement governing the work, the Earl was to receive 95,000 acres (35,450 hectares) of reclaimed land and the Crown 12,000 acres, with the revenue from 40,000 acres of the Earl's land to be used to maintain the works. Thirteen other 'Gentlemen Investors' joined the Earl in this enterprise.

The Adventurers appointed Sir Connelius Vermuyden as Director of Works. He had already established himself in England by draining Hatfield Chase, for which Charles I honoured him with a knighthood, and which may also have convinced Charles that drainage of the Great Level of the Fens could be accomplished. The Lynn Law Agreement contained no mention of having to render the land permanently dry and indeed left rather vague the degree of drainage which had to be achieved. By 1637 Vermuyden had completed the work to the satisfaction of the Commissioners of Sewers and they recommended that the Earl's 95,000 acres should be allotted to him. This met with opposition from the Bishop of Ely as well as other inhabitants of the Fens and possibly Charles himself, which resulted in a reversal of the decision in 1638 and the work declared incomplete and defective. Charles then assumed the role of Undertaker and reappointed Vermuyden, who managed to complete only some minor works before the differences already sprouting between Charles and the Commons grew rapidly into an enmeshing thorn bush of civil war on which the country lacerated itself between 1642 and 1648. Charles' interest in draining the Fens ceased when he lost his head in 1649, but his conqueror, Oliver Cromwell, himself a Fenman and greatly assisted by Fen people in his defeat of the King, supported the so-called 'Pretented Act' passed by the Commons in 1649 for 'drayninge of the Great Levell of the Fennes'. On 4 June 1649, fourteen Adventurers led by William, Earl of Bedford (his father Francis having died in 1638), and the Earl of Arundel met and decided to approach Vermuyden to ask him to become Director of Works on agreed terms. Vermuyden eventually agreed to take up this position, but only after considerable haggling over the terms of his appointment, and after the Adventurers had approached others to submit proposals for the work, including his fellow Dutchman and arch rival Westerdyke, who advocated the embanking of existing rivers, in direct contrast to Vermuyden's preference for digging straight lengths of channel to cut off meandering lengths of river and thus increase the flow velocity and self-scouring action of the river.

In his earlier work on the Great Level, Vermuyden had dug a straight channel that came to be known as the Old Bedford River, 70 feet (33 metres) wide and 21 miles long from Earith to Denver, cutting off a tortuous 30-mile length of the River Ouse passing through Ely. Sluices built at each end of the cut controlled inflow into it and kept tidal water out. The purpose of this cut was to take the water flowing into the upper reaches of the River Great Ouse from Bedfordshire and Huntingdonshire, but other tributaries

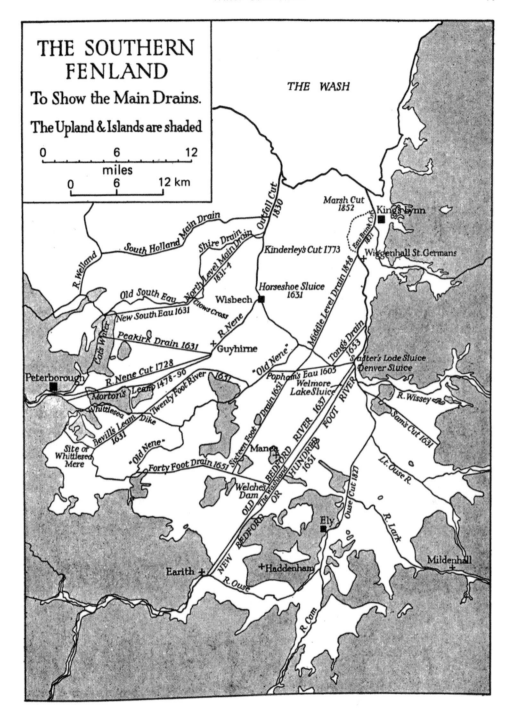

Drainage system of the English Fenland in 1938, showing approximate dates for completion of artificial cuts.

Driven flush with ground level in 1851, the top of this Holme Fen Post now stands some 4 metres above ground level as a result of land drainage. As seen in this photograph taken in 1932, 3.25 metres of shrinkage occurred in the first 81 years and some 0.75 metres has occurred in the 81 years since then.

downstream of Earith – the Cam, the Lark, the Little Ouse and the Wissey – still flowed into the Ouse. The work at this time also included a number of new auxiliary drains and the enlargement of existing drains including Morton's Leam.

Vermuyden's mind had been active during the Civil War on devising a plan for the proper drainage of the Great Level, which comprised for the most part the peat fens and which he divided into the North, Middle and South Levels, repairing the works of earlier drainers, raising new banks, cutting new channels including the Forty Foot Drain from Ramsey to discharge into the Old Bedford River, and constructing sluices. His major work, however, came in the South Level with the construction of the New Bedford River, also known as the Hundred Foot Cut from its width, parallel to, and about half a mile to the east of, the Old Bedford River, which became the main channel for the waters of the Great Ouse. The Nine Holes Sluice taking water into the Old Bedford River from the Great Ouse remained closed except in times of flood. In order to cope with flood water, the outer banks of the Old and New Bedford Rivers were made higher than the inner banks, so that the flood waters admitted through the sluice into the Old Bedford River and the floodwaters in the New Bedford River spilled over the inner banks and inundated the land between the two, thus confining the water and stopping it flooding the valuable farmlands. This system of flooding the Hundred Foot or Ouse Washlands, as this area is now called, is still used today to prevent floodwaters from the upper reaches of the Great Ouse River reaching the Fens.

It says much for the tenacity of Vermuyden and his engineers that they completed these substantial works using only manpower and horses, shovels, wheelbarrows and carts, under constant threats and harassments by local inhabitants and having to cope with financial uncertainties. Some good fortune, however, attended their efforts, as Cromwell's victory over the Scots at the Battle of Dunbar provided a ready supply of Scottish prisoners of war to do much of the digging. A further supply of prisoners of war, this time Dutch, became available later in the work after a naval engagement in the Channel.

Maintenance and reconstruction of drainage works in the Fens is a job that is never completed and indeed to some degree is self-defeating. Lowering of the water table in the peatlands increases the effective overburden stress in the peat and causes extreme settlements. Added to this, the surface peat dries out, oxidises and blows away as dust. There is evidence that some areas of the fens have experienced a drop of three metres and more in surface level over the past 150 years.

4

Water Transportation

Ports and Harbours

Islamic sea-going trading routes more than matched their long-distance overland pack-animal routes. They established shipping routes across the Caspian Sea and Persian Gulf, sailed the Indian Ocean to India and China and for a time their trading ships, with their distinctive triangular lateen sails, dominated much of the Mediterranean under the protection of powerful naval forces. And they had outstanding goods to trade, created by skilled craftsmen in the great cities: fine steel from Damascus, also known for its woven figured silk or 'damask'; silks, jewellery, glassware and pottery from Baghdad; leather from Morocco; swords from Toledo; cotton cloth from Mosul. The trading ships and naval forces required port and dockyard facilities, and while captured Byzantine facilities were pressed into service these soon proved inadequate, leading to the need for new shipyard constructions – for the eastern trade at Basra on the Shatt al-Arab, at Siraf in the Gulf and Suhar in Oman, and for the Mediterranean, in order to maintain naval dominance, upgraded or newly constructed docking and harbour facilities stretched from Acre and Tyre in the east; Rawda (Cairo), Alexandria, Tripoli, Mahdiyya and Tunis in North Africa; Seville, Almeria and Valencia in Spain; and Messina and Palermo in Sicily. Excavation of a hillside at Tunis gave accommodation for 200 ships and at Mahdiyya, also on the Tunisian coast, excavation into a rock face provided anchorage for thirty large ships.

The powerful Maritime Republics of Italy created thriving ports such as Genoa, Pisa, Amalfi and above all – as depicted vividly by Canaletto – Venice, which dominated trade in the Mediterranean between 1050 and 1300, having taken over from Islam the former Byzantine routes. They prospered, too, from the eastern depredations of the Crusaders. Their control in the eastern Mediterranean, where both the Genoese and Venetians undertook new port constructions, resulted in the import into western Europe of luxury items such as perfumes, gems, spices and fine textiles in exchange for agricultural goods. In Genoa itself, its exposed east coast was protected by the construction of a breakwater with a walled superstructure to serve, additionally, as a fortification. In 1245, in recognition of its importance to the city, the authorities designated this a 'Pious Work', which imposed upon the citizens the mandatory requirement to include in their will a legacy to support maintenance of the breakwater.

Siltation posed a constant problem and in many cases led to the need, from time to time, to evacuate the harbour basin and dry it out so that the accumulated silt could be removed. Efforts to try and avoid this costly and disruptive work led to experimentation with, and in some cases the implementation of, innovative solutions, particularly in Italy in the fifteenth and sixteenth centuries. Francesco di Giorgio Mantini designed a dredger consisting of twin hulls supporting between them a wheel with four arms having buckets at their extremities. Leonardo sketched a similar design, as did Veranzio, although in the latter case the revolving wheel was replaced by a grab mechanism formed by two opposed matching buckets. Other methods used to remove the silt included creating water currents within the basin to sweep it out, such as opening sluice gates in the moles enclosing the basin or lowering wooden plates hinged to the sea bottom at the harbour entrance.

Organised trade in northern Europe, which largely died immediately following the Roman period, started to flourish again in the twelfth and thirteenth centuries from centres such as Visby on the Swedish island of Gotland, which dominated the Baltic and traded as far east as Novgorod in Russia. In 1241 the German port of Lübeck on the Baltic and its near neighbour Hamburg formed an alliance to monopolise the salt-fish trade in the Baltic and North Sea. This alliance became the genesis of the Hanseatic League, which at its height in the fourteenth and fifteenth centuries embraced the merchant guilds of perhaps 160 sea-port and inland towns stretching from Novgorod to London and beyond, with principal Kontors (outposts) at Bergen, Bruges and London. The German Hansa merchants from Lübeck, Hamburg and Cologne based in London established in 1320 a walled community called the Steelyard on the north bank of the Thames (now the site of Cannon Street station), just west of London Bridge, near the home of customs officer Geoffrey Chaucer, with its own warehouses, counting houses, weighing houses, church and residential quarters. The League also established an important Kontor in King's Lynn, the medieval warehouse of which still stands and is now the town's register office. Over a century after the alliance of Lübeck and Hamburg, during which time other towns joined this association, the Hansa town representatives held a Diet in Lübeck in 1361 which is usually considered to mark the formal establishment of the Hanseatic League and the recognition of Lübeck as its chief town. It is still referred to as a Hanseatic town today and is recognised as a world heritage site by UNESCO.

The Hansa merchants dealt primarily in bulk products such as fish, grain, beer, salt, timber, furs, wool, cloth, iron, lead and pitch. Fish from Norway, grain from Poland, furs from Russia, beer from Germany, salt from Spain and Portugal, timber from Germany and Prussia, wool, lead and grain from England and cloth from Flanders. Bruges became the second most important Hansa port, partly because of the Flanders cloth industry and also because this is where trading vessels from the Mediterranean and beyond exchanged goods with the Hansa merchants. The alliance reached its peak in the fourteenth century, during which it dominated trade in Northern Europe, but by the sixteenth century it was in decline and, although never formally dissolved, the last Diet held in 1669 attracted only nine members. Factors contributing to the decline included internal dissension and increasing competition from local traders, with individual towns and countries putting their own interests before those of the Hansa merchants. In 1487 Tsar Ivan III captured

Novgorod and expelled the German merchants. In the sixteenth century, the silting up of its shipping channel eventually made Bruges inaccessible to the larger Hanseatic ships. Responding to pressure from local merchants, Queen Elizabeth closed the London Steelyard in 1598 and, although reopened by King James, it no longer represented a threat to local English traders.

Archaeological excavations of medieval ports in London have revealed them to have consisted of timber revetments built out in the water some distance from the foreshore with the intervening area filled with refuse, surfaced with gravel or stone. As well as providing a docking facility and disposing of unwanted refuse, this method of construction led to a valuable increase in land area. Developments of this form of construction included a double row of revetments, the front row permanently submerged and its top separated by horizontal timbering from a higher landward row, to counter problems of silting and provide a deep water berth. English landowners and English ports grew rich on the export of wool, and increasing trade demanded ever bigger ships and better ports and port facilities. Beaching and careening these ships for repair was no longer possible and dry docks of timber or masonry had to be provided. By 1500 dry docks had become a common feature of port facilities.

In the first half of the seventeenth century the Dutch led the maritime nations in Europe, commanding no less than two-thirds of the trade in 1650. But it didn't last. Their landlocked harbours needed constant dredging and the pontoon-type ladder dredges used, operated by four horses, could only dredge to about 5 metres depth, which limited the draft of the Dutch vessels. The English and French found that speed and sailing power improved with a deeper draft of about 6 metres and with their harbours able to cope with the deeper draft, they put the Dutch out of business in 40 years.

The Atlantic countries of England, France, Portugal and Spain supplanted the great Italian ports of the medieval and Renaissance periods and competed fiercely among themselves to exploit newly discovered, far distant lands. The achievements of the Elizabethan Age owed as much to the engineers who created the great port facilities of Plymouth, Southampton, Portsmouth, London and Bristol as it did to the remarkable men who sailed from these ports – men like Hawkins, Drake and Raleigh.

Competition between France and England spurred François I, contemporary and rival of Henry VIII, to commission a harbour to be built at Le Havre ('The Port') at the mouth of the Seine. Completed at the end of the sixteenth century, jetties, lock gates and basins were built with protective measures to withstand a tidal range of 8 metres. Stone jetties 8 metres high, laid in cement and held together with iron clamps, extended 30 metres into the sea. Well aware that a port serving not only Paris but the whole nation, as well as providing a bastion against the English, needed people as well as jetties, lock gates and related structures, François proclaimed on 8 October 1517 that for ten years all who lived within the district would be free of taxes. They would also enjoy free salt for their fishing and personal use. He commissioned his engineer le Roy to bring fresh water by clay pipes from a spring 5 km distant.

Although known mostly for his work on fortresses and fortifications, the outstanding French engineer Vauban concerned himself with a range of construction works, including harbours. Notable among these were the extensive harbour and coastal works carried out in the seventeenth century as part of the work to convert Dunkirk into an impregnable coastal fortress. The works included canals, river basins, two long jetties flanking the entrance channel, storehouses and workshops. Overshadowing all the work, however, was the great lock marking the entrance to the inner harbour, the construction of which Vauban supervised himself. In using locks at several of their harbour works, including Le Havre and Cherbourg as well as Dunkirk, to cope with changes in tide levels, French engineers introduced a technique first used by the Chinese to take up changing levels in the construction of their canals. The locks retained adequate water depth at low tide level to float their sea-going vessels. They also provided a convenient means of cleansing the basins by opening the lock gates and allowing the water to rush out, taking the silt and debris with it.

Improvements to the port of Dunkirk were made in 1715 by another outstanding French civil engineer, Belidor, who built locks to provide wet dock facilities. Dunkirk remains an important port to this day, serving the highly industrialised Pas-de-Calais area of northern France. In his book *Architecture Hydraulique*, published between 1737 and 1753 in four volumes, Belidor describes various aspects of harbour design and construction, including dredging methods and dry dock construction. Recognising the great difficulty in rendering dry docks watertight, he recommends keeping the floor as high as possible and even adopting a two-chamber installation, pumping water in to raise the ship from the first chamber to a higher level in the second chamber.

With all this activity in Europe and particularly in France, Britain could not stand still, and expended considerable effort and money in the seventeenth and eighteenth centuries upgrading existing ports or building new ports to accommodate the largest ships. The Howland wet dock on the Thames, built in 1703, boasted two dry docks, the largest of which measured 75 metres in length, 13.5 metres in width and 5 metres in depth at high tide. It could accommodate the largest merchantmen of the time. The scene was being set for the great rivalries, civil and military, which dominated relationships between these two leading powers in the eighteenth and nineteenth centuries. Civil engineering was one of the areas in which this rivalry manifested itself most strongly.

Canal and River Navigations

It may be felt that as a result of his unsolicited and unwelcome contacts with Islam, some of the Moorish expertise in hydraulic engineering might have rubbed off on Charlemagne. Unfortunately it did not. If it had, he might have avoided the embarrassment of commencing work on a canal in 793 AD to link the rivers Regnitz and Altmühl and so allow passage from the Danube to the Rhine. Although he mustered a substantial labour force, and the required cut amounted to only 2,000 paces in length and 9 metres in width, the project ground to a halt because he failed to overcome the difficulties of excavating through marshy ground, beset by high ground-water levels, slumping banks and liquidising of excavated soil.

Civil engineering activity in Europe relating to inland navigation, from the time of Charlemagne through to the fifteenth century, consisted mostly of minor works, maintaining the navigability of the major rivers. Some stretches of the rivers experienced heavy usage, despite the imposition of tolls by the local lords: in the fourteenth century, for example, passengers and light goods moved regularly between Koblenz and Andernach, Mainz and Oppenheim, Mainz and Frankfurt. Traders known as Easterlings (the origin of the word Sterling for British currency) travelled by river to the burgeoning cities and international fairs in Europe, eventually contributing to the rise of the Hanseatic League. The Paris-based Hansa merchants obtained concessions from the King to navigate the Seine, to build a port at Bercy above Paris and to charge tolls.

Netherlands civil engineers established their premier position, retained to this day, as outstanding hydraulic engineers as a result of their work to maintain and improve the navigability of their inland waterways, along which the Baltic trade flourished. Cheese, fish, salt and manufactured goods flowed to the Baltic in exchange for timber, grain, furs and pitch. Stanches with a single vertically rising gate controlled water levels and came into common use in Europe.

Mill-dams built to impound water to drive water wheels presented an obstacle to medieval river traffic in Europe and led to various techniques being devised to avoid offloading and reloading of goods to transfer them past these points. Inclined planes or slipways, over which the boats could be hauled by windlass-assisted manpower or horse-power, provided one solution, but a more satisfactory technique employed some form of gate built into the mill weirs. Until the invention of the pound lock, these gates operated as simple flash-locks. One form of gate consisted of vertical posts attached at their tops and bottoms to horizontal beams, the top beam being extended to balance the weight of the gate, thus enabling the gate to be rotated about a support point on the beam. Grooves in the vertical posts received planks that could be lowered into place by long handles or ropes. Normally the planks were in place to provide an adequately water-tight barrier, but when a boat wished to pass they were lifted out and the framework of the gate, consisting of vertical posts and horizontal beams, rotated to one side to leave a clear opening.

In Europe, tidal chambers to overcome differences in tide and river levels preceded the use of the pound lock, which, in contrast, served the primary purpose of overcoming differences in land levels. Nevertheless, they operated on a similar principle and it is likely that experience with tidal chambers was the key to the introduction of the pound lock in Europe. It cannot be ruled out, however, that knowledge of the pound lock found its way into Europe from China through the medium of visits to China by Europeans pursuing commercial and trading interests there. Marco Polo was by no means the only European of the time to visit China, the uniqueness of his visit being simply that he remained in China for seventeen years in the employ of the Kublai Khan, and on his return wrote a detailed account of the country. He travelled to China in the company of his father and uncle, both of whom had been to China previously.

The first example of a locked tidal basin dates from 1373 at Vreeswijk in the Netherlands, its purpose being to allow boats to enter the River Lek from the Utrecht canal. As the basin had adequate capacity to accommodate several vessels, it sufficed to operate the lock only three times a week, at two o'clock in the afternoon. No doubt the local townsfolk and boatmen mingled to their mutual advantage and pleasure during this enforced stopover. Several similar Dutch tidal basins followed in the fourteenth century, but particular mention must be made of a lock at Damme, near Bruges, that with masonry side walls 30 metres long and 10 metres apart, was too small for a holding basin and may have been operated whenever a boat arrived there. In 1405 Bruges lost the advantage derived from this lock, as a great storm initiated a process which ultimately led to the silting up of the approach channel to Damme, making it unreachable. Bruges consequently lost its advantage to Antwerp.

Towards the end of the fourteenth century the Duke of Saxony, or at least his advisers, recognised the benefits that would accrue from linking by an improved inland waterway Lübeck and Hamburg, the two great towns of the Hanseatic League. Lübeck was situated on the River Trave, and just upstream from the town a tributary, the Strecknitz, which ran due south to Mölln Lake, some 30 km from Lübeck, had been made navigable earlier in the fourteenth century. Sixty kilometres to the south of Lübeck, the town of Lauenburg could be reached by navigation from Hamburg along the River Elbe. In 1390 the Duke agreed that the River Delvenau, flowing into the Elbe at Lauenburg from the north, should be made navigable, an enterprise that required the provision of eight flash locks. Access to Lake Mölln from the River Delvenau required a canal running initially horizontally from the river for a distance of 11 km in a cutting up to 3.5 metres deep, then dropping 5 metres in a distance of three-quarters of a kilometre to the lake. As the water supply for this section of the canal depended upon ground water seeping into it from the sandy soil, the large water losses attendant on the use of flash locks would have severely limited the frequency of boat passages through the canal. Flash locks may have been used originally, in which case sheer frustration or more likely loss of revenue from tolls led to their demise, to be replaced by two pound locks. Probably in the interests of making maximum use of the limited water available, the wooden chambers were made sufficiently large to accommodate ten small barges, each 11 metres long and 3.3 metres beam, and, as with the tide lock at Vreeswijk, the gates were operated only every second or third day. In consequence, the passage from Lübeck to Lauenburg took eight to ten days; despite this the system, with some changes and improvements, remained in use until 1900, when the present Elbe–Trave canal superseded it.

In 1385 Galeazzo Visconti decided that his rise to power in Milan called for the construction in that city of a cathedral more glorious than any other in Europe. Marble for the construction of this great edifice was available from the quarries of Candoglio on the western side of Lake Maggiore, and the obvious means of transport was by water across Lake Maggiore, down the Ticino River to the offtake point of the Naviglio Grande canal, then to follow this canal to Milan. The transportation of the vast amount of stone required for the cathedral clearly meant disruption to the canal's original and primary use for irrigation and lesser use for operating water mills, but the farmers and mill owners had no choice but to bow to the power of the Church and City State. The position worsened further when merchants quickly realised the possibilities of using the canal for transporting their goods now that the precedent had been set. It is of interest that today the primary use of the canal is again irrigation, as the need for inland navigation in the Po Valley is limited.

Initially the stone was offloaded at the edge of the Darsena, an artificial basin on the outskirts of Milan, where the Naviglio Grande terminated; from here the stone had to be trundled to the cathedral site through the already congested roads. A moat supplied by water from a small local stream already surrounded the city walls, with a connection to a small basin near the cathedral site, and it was decided to utilise this for the transport of the stone. Direct communication from the Darsena basin to the moat presented a problem, with the moat having a level about 3 metres higher than that in the Naviglio Grande. A channel excavated to link the two initially consisted of a single barrier of stop planks, and the passage of boats was effected by closing all irrigation gates and overflow devices in the Naviglio Grande and at the same time depressing the water level in the moat. When the two levels corresponded, removal of the planks allowed the boats to make the passage. After replacing the planks, the moat level again rose and the boats could proceed to the site. A similar unwieldy and lengthy procedure allowed the empty boats back into the canal.

It is surprising that this procedure, which caused great inconvenience to all users of the canal, continued for some decades before the implementation of the simple solution, in 1438, of providing a second barrier close to the first, and thus creating a pound lock. Contrary to the once-popular belief that Leonardo da Vinci invented the first pound lock in Europe, arising from the fact that his sketches included representations of such locks, this Milan lock, known as the Viarenna Lock, predated Leonardo's birth by fourteen years. Although it was the first pound lock in Italy it had, as already seen, predecessors in Europe outside Italy; but it certainly triggered a spate of locked canal construction, amounting to some 90 km of canals with twenty-five locks serving the city of Milan alone by 1475. Leon Battista Alberti gave the first known description of a pound lock in his treatise on building, *De re Aedificatoria*. Manuscript copies of the book existed as early as 1452, the year of Leonardo's birth, but it was not printed and published until 1485 and became the first ever book on construction to be produced by the process of movable type printing introduced in Europe around the middle of the sixteenth century. His description of the lock itself is clear, although he omits to mention the most fundamental principle of the lock – the equalisation of water levels.

> To each lock you ought to make two stops, cutting the river in two places, and leaving a space between them equal to the length of a vessel, to the intent, that if the vessel is to ascend, when it comes to the stop the lower sluice may be shut the upper one opened; or if it be to descend the upper one may be shut and the lower opened...

Immediately before this passage, he describes a sluice gate with a flap-valve in the middle of it, turning about a vertical spindle, but with unequal widths on either side of the spindle. The unequal pressures of water on either side of the spindle opened the valve and, according to Alberti, could be opened by a child. The flap valve allowed equalisation of water levels on either side of the sluice gate to facilitate its opening. Leonardo sketched a flap-valve identical to that described by Alberti and in doing so simply recorded a device already in quite common use.

In 1451 Bertola da Novate took over the post of ducal engineer of Milan, a position later occupied by Leonardo. He became unquestionably the outstanding canal engineer of the fifteenth century. He undertook as his first major task to link Milan by canal to Pavia, some 35 km to the south, near the junction of the Ticino and Po rivers. Topographical features dictated against the construction of a direct canal and Bertola decided to extend the Naviglio Grande southwards from Abbiate to terminate at Bereguardo, a small town some 11 km from Pavia. To cope with the drop of 25 metres in its length of 19 km, Bertola installed eighteen pound locks to control such a steep gradient on a major canal. Existing locks on the canal, which is still in use today, probably occupy the sites of the original locks.

Francesco Sforza's ambitious plans for a vast network of canals over northern Italy, to extend the commerce of Milan, kept Bertola busy, on top of which he carried out some permitted freelance work for other ruling princes, notably at Mantua and Parma. His next major work for the Duke of Milan, following immediately on the completion of the Bereguardo Canal, was the construction of the Martesana Canal to link Milan to the Adda River east of the city. This multi-purpose canal offered many potential benefits, including direct communication with Lake Lecco in the north, a branch of Lake Como from which the Adda flowed, and gave an additional route to the Po valley in the south; irrigation water and water-power could also be derived from the canal. In fact, Bertola did not achieve the navigable connection with the northern lakes because of

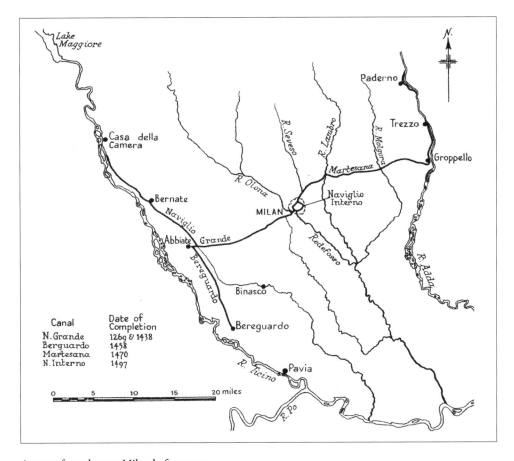

A map of canals near Milan before 1500.

the exorbitant expenditure that would have been required to provide locks to overcome the 60 metre drop in the Adda River from Lake Lecco to the canal take-off structure at Trezzo, 30 km to the south. A further attempt to complete this link by means of a lateral canal (a canal which parallels the course of a river) with ten locks was commissioned in 1518 by Francis I of France, who then ruled Milan, but this work, designed by Benedetto Missaglia, came to a stop when Milan fell to the Holy Roman Emperor Charles V. Nevertheless the Martesana Canal, completed by Bertola between 1452 and 1470 and linking Milan to the Adda constituted a major accomplishment. At Trezzo a large weir raised the water level for the canal take-off; the canal then followed close to the river for a distance of 8 km, separated from it by a substantial masonry wall still in existence today, then turned westward to cross the Lombardy plain to Milan. The 38-km-long canal had to cross a river, the Malgora, and a smaller stream, the Lambro, the former by means of the first known three-arch masonry aqueduct and the latter by taking the stream under the canal in a culvert. Two locks sufficed along the canal and at the Milan end the boats entered a basin or dock near San Marco.

Leonardo da Vinci succeeded to the post of ducal engineer in Milan in 1482 at the age of 30 and one of the tasks given to him by Ludovico Sforza was to connect the Martesana and Naviglio canals through the internal waterways of Milan, the Naviglio Interna. He had six new locks constructed in this internal system; but, more notably, in 1497 he constructed a lock at San Marco to link the terminal basin of the Martesana canal to the Naviglio Interna. This lock features in Leonardo's celebrated sketches, which illustrate the use, possibly for the first time, of mitre gates, which, by virtue of forming a vee upstream, are held tightly closed and watertight by the pressure of the water. A simply operated flap-valve, already in common use, is shown fitted into the gates to allow equalisation of water levels before operating the gates. This first completely modern form of lock, rectangular in plan, had a length of 29 metres and a width of 5½ metres. A similar lock sketched by Leonardo shows a pile cut-off below the masonry walls.

Leonardo's personal interests may have been directed more to river training than canal engineering, a legacy of his early life spent in the workshop of Verrocchio in Florence, close to the turbulent River Arno. According to Vasari, while still young and studying in the workshop of Andrea del Verrocchio in Florence, he proposed a scheme for transforming the Arno from Pisa to Florence into a navigable stream, making Florence an important inland port with its own direct waterborne trading links with the outside world. This possibility had no doubt long occupied the minds of prominent citizens of the city and in 1487 Luca Fancelli, a leading engineer and architect, probably responding to this interest, wrote to Lorenzo de Medici with a proposal for canalising the Arno. Interestingly, Luca and Leonardo were both living in Milan at this time, in the employ of the House of Sforza, working together on Milan Cathedral and perhaps other projects, and they may well have discussed the possibilities of controlling the Arno. Luca had little experience in major hydraulic works and it is possible that he simply took the opportunity to pick Leonardo's brain before writing to Lorenzo with his proposal. Whatever the truth, Lorenzo did not invite him to submit a design for improving the river. Lorenzo died in 1492 and the upheavals in Florence over the next few years, with the ascetic monk Savanorola effectively ruling the city, did not present conditions conducive to conducting such works.

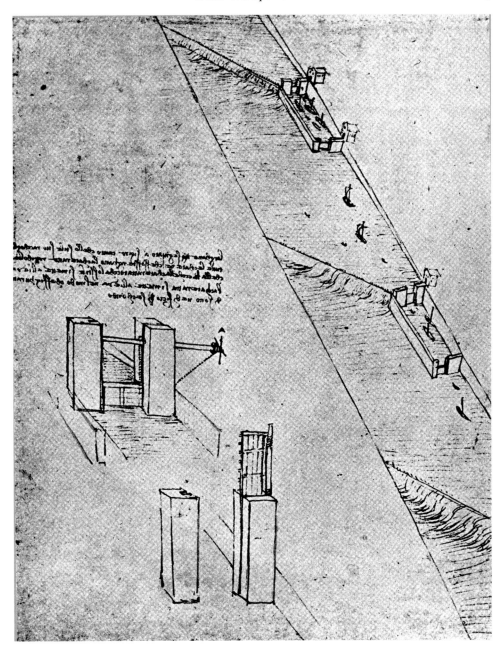

Leonardo's sketch of locks (with boats shown) and details of lock gates.

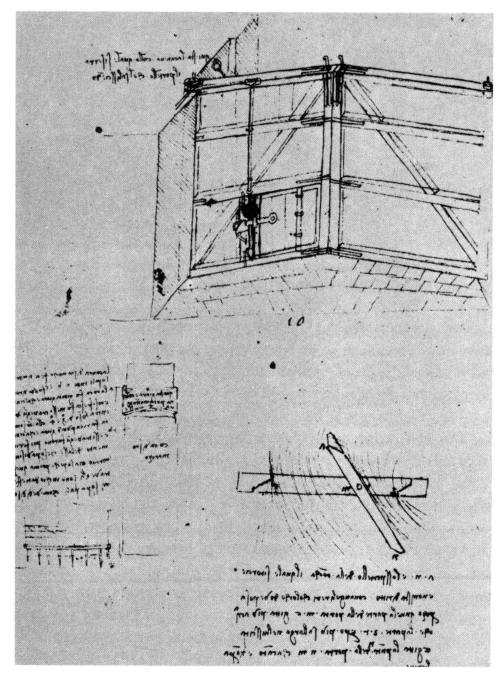

Leonardo's sketch of mitred lock gates with a wicket to admit water from the higher level. Each hinged gate folds back into a recess in the wall, as depicted in the smaller sketch.

Leonardo remained in Milan until 1499, when the French under Louis XII captured the city, causing his patron Ludovico Sforza to flee. During this seventeen-year stay in Milan control of the Arno continued to occupy his mind, along with his many other interests. He saw the project as multi-purpose, to provide a navigable waterway, to control flooding and to provide both irrigation water and power generation. As a means of controlling flooding and regulating water flow, he proposed damming the river at a point upstream of Florence and, rather than canalising the river itself, he envisaged a canal leaving the Arno close to Florence and following a great northerly loop through important towns such as Prato and Pistoia to rejoin the Arno before it reached Pisa. Despite diverging as much as 30 km to the north of the Arno, he claimed the route would be shorter than the river because of its sinuosity. At this time, without having had the opportunity of visiting the route, he had in mind a fully excavated canal without locks, because they had limited life and imposed undue costs and maintenance. This would have required several deep excavations, including one of 69 metres depth through a ridge known as Serraville, through which a railway tunnel now passes.

Taking the precaution of sending 600 gold sovereigns to be deposited in his account in Florence, Leonardo made brief visits to Mantua and Venice before fetching up again in that city, where in 1502 he received a letter from Cesare Borgia inviting him to be his chief architect and military engineer, based in Imola. It happened that also in temporary residence in Imola at this time was Niccolò Machiavelli, second Chancellor of the Government (Signoria) of Florence and responsible for Florentine military and foreign policy. His brief as emissary was to report to Florence on the intentions and activities of Cesare Borgia. There is no direct evidence that the two men met during this time, but they must certainly have done so and it is equally certain that Leonardo took the opportunity to impress upon Machiavelli the advantages to Florence that would accrue from implementation of his long-dreamed of project of a navigable waterway linking the city to the Ligurian Sea. Machiavelli leant a ready ear to this proposal, not least because he realised how, by an extension of the project, Pisa, the great trading and military rival to Florence, could be landlocked and starved into submission by changing the course of the Arno, so that it flowed directly into the sea at Livorno.

Leonardo returned to Florence in 1503 and with the encouragement of Machiavelli, also now back in Florence, set about preparing a more detailed plan for a canal. With the prospect of the work actually going ahead, he abandoned the idea of making deep cuttings and planned a canal following his proposed northerly route with locks, at least one major tunnel and a three-arched masonry aqueduct to cross the Bisenzio river, with water for the canal to be supplied not from the Arno, but from the Bisenzio and the Ombrone rivers. Under the protection of the Florentine army, work started in 1504 on digging huge ditches to divert the mouth of the Arno away from Pisa, but project difficulties arising from high costs escalated into disaster, caused in part by violent storms that flooded the whole plain and collapsed the walls of the deep excavations. Boats guarding the mouths of the ditches sank, costing many lives. The whole project was abandoned after about two months, never to be resumed: an outcome not unfamiliar to Leonardo. In order to help Leonardo financially at this time, Machiavelli contrived for Leonardo to be commissioned to paint a large fresco on the wall of the Grand Council

Hall of the Palazzio Vecchio to celebrate the victory of Florence over the Milanese, about which Leonardo may have had mixed feelings – *The Battle of Anghiari* – but this fared no better than the canal project, Leonardo having adopted an untested technique that caused the paint to drip and run. In 1509 hostilities ceased with the defeat of Pisa, in which Machiavelli played no small part, giving Florence full control on the Arno and burying any further thought of constructing a canal at great cost. Leonardo returned to Milan in 1506.

Leonardo's writings, like his sketches, contain much informative detail. Among his notes in the Codex Atlanticus, estimates are set out of the manual work input and costs of canal construction. He is simply recording computations of a sort that all canal engineers at this time would have made. They, no less than today's civil engineer, must have striven to keep costs to a minimum and Leonardo, for example, observed that the best time for digging a canal fell between the middle of March and the middle of June, when peasant labour was cheapest. Leonardo also sketched machines suitable for canal excavation, again to reduce the cost of human labour.

In 1516, when Leonardo crossed the Alps to join the King of France, François I, he still retained his life-long love of river and canal engineering. Installed comfortably by the king in a chateau in Amboise, and living on water-rights accruing from the canal of San Cristofano in Milan on which he had worked, he spent his last days until his death in 1519 reflecting on matters of particular interest to him, and on his incessant sketching and writing. Most notably, he proposed to the king a watershed canal connecting the Mediterranean and the Atlantic, either by joining the rivers Saone and Loire or the Garonne and Aude. Circumstances were not yet right for this visionary proposal to be implemented, but the Languedoc Canal, based on the River Garonne, became a reality 150 years later, becoming the first canal in Europe that could claim parity with the Grand Canal of China.

Throughout the sixteenth century the plains of Lombardy remained the centre of canal engineering in Europe. Any challenge that might have emanated from France disappeared when, after the death of François I in 1547, a series of religious wars convulsed the country, pitting rebellious Protestants against the Catholic royal authority of Henry II, who succeeded François I and, after his death in 1559, of his Italian wife Catherine de Medici, who acted as Regent during the reign of Henry's sons.

When a serious flood damaged the medieval intake of the Naviglio Grande from the Ticino in 1584, Guiseppe Meda, a painter turned civil engineer, undertook the urgent task of its reconstruction. Working under great pressure, as all traffic on the canal had been stopped and the mills along its length could no longer generate power, he replaced the medieval stone breakwater destroyed by the flood, which had divided the river longitudinally, with a weir across the river and control works to regulate the flow of water and maintain it at a constant level. This highly successful solution boosted Meda's reputation and led to other commissions.

Over a hundred years after the completion of Bertola's Martesana Canal from Milan to the Adda River, the problem of making navigable the steep section of the Adda above the Martesana intake had still not been solved. Missaglia's work on providing a by-pass or lateral canal with ten locks had ceased in 1522 by order of Charles V. Meda drew up plans to shorten the length of the canal to bypass only the steepest section of the river and to take up the 27 metre drop with two locks of unprecedented heights of 18 metres and 9 metres, appropriately known as shaft locks. Recognising that gates of these heights could not be made completely watertight, he made provision for leakage through them. On completion of the canal, he was to receive from the city of Milan two-thirds of the river tolls for himself, his heirs and assigns in perpetuity, in addition to the right to build wharves and mills on the banks of the river wherever he wished.

After many years of negotiations, financial intrigues and delays, the work started in 1591, but it soon became plagued with problems: last minute changes, masonry walls damaged by frost, worker insubordination and the city's insistence on constructing a towpath along the river, all of which interfered with the work. Eventually Meda was charged with incompetence and sent to gaol. He died penniless shortly after his release in 1599. His demise marked the end of major canal building in Lombardy.

Just as the end of the sixteenth century saw the demise of canal building in Lombardy, it also saw it take off in France following the cessation of over thirty years of Religious Wars– Catholic Guises against Catholic Montmorencys, Huguenot Protestants against Catholics, Calvanists against Guises – which had ravaged the country. In 1598 Henry IV, having abjured his Protestantism to become King, declared a general amnesty in France and issued the edict of Nantes, granting religious and political freedom to all his subjects. He set to work with his Minister of Finance and Grand Voyer (Director of Communications), Maximilien de Bethune, Baron de Rosny and now better known as the Duc de Sully, a Protestant, to expand the commerce of France and rebuild the devastated cities. The joining of the Atlantic to the Mediterranean had long been mooted by Leonardo and others, but Sully sensibly gave priority to improving the internal economy of the country by linking Paris to the provinces by waterways. At Sully's instigation, the contractor Hugues Cosnier drew up a proposal to link the Loire to the Seine and hence to Paris.

Cosnier planned originally to make navigable the Trezée River, which flowed in a south-westerly direction into the Loire at Briare and the Loing River, which flowed in a northerly direction into the Seine, and to join the upper waters of these two tributaries with a summit canal. The 38 metre ascent from Briare to the summit and the 79 metre drop to Montargis, after which the Loing was already navigable, was to be accomplished with forty-eight timber locks. The work started in 1605, but within a year it became clear that river flooding would be a problem during and after construction, and timber locks would be inadequate and require constant maintenance. Under a new contract arrangement, masonry walls nearly 2 metres thick, to resist heavy earth pressures, replaced timber for the locks and an all-excavated canal solution was adopted. The route followed the Trezée valley some 11 km from Briare, then crossed the plateau separating the Trezée and Loing rivers in a northerly direction, with a summit canal 6 metres long descending the steep sided valley of the Loing to Rogny in a unique staircase of six locks with a total drop of 20 metres; after this it followed closely the Loing River for 38 km to Montargis. Water for the summit section of the canal, to feed the locks, came from the upper Trezée, through a lake, the Etang de la Gazonne, which acted as a reservoir; as an additional precaution a 2.8 km length of the summit canal was deepened to provide an intermediate lock and give additional storage capacity.

Sully supplied 6,000 troops to provide Cosnier with a reliable workforce and took great interest in the work, as did Henry IV, who honoured Cosnier by visiting the site in 1608; but this royal patronage ended abruptly on the death of Henry at the hands of an assassin in 1610. His son and successor Louis XIII, only eight years old at the time of his accession, ruled for several years with his mother Marie de Medici as Regent. She had little enthusiasm for Sully and his works and readily accepted the findings of a Board appointed to review the project, that it should be abandoned on the grounds that its maintenance costs after completion would be prohibitive.

By 1611, when work stopped, Cosnier had completed three-quarters of the project. He deserves a place among the highest echelons of outstanding French civil engineers. His work lay abandoned and unattended for twenty-seven years when Louis XIII, at the strong urging of the enlightened Cardinal Richelieu, granted letters patent in 1638 to promoters Jacques Guyon and Guillaume Bonterone to finish the work in four years and pay the outstanding land debts in exchange for ownership of the canal. Modifications made to the design of the canal included decreasing the number of locks to forty-one, reducing lock lengths from 42.5 metres to 33 metres and providing a second source of feed water from the Loing River to the summit to operate the Rogny staircase. The canal width was fixed at 4.25 metres, and depth at 1.6 metres.

The promoters had every reason to be pleased with their investment, which, mainly through the transport of coal and wine to Paris, gave an annual return of 13 per cent on capital in the second half of the seventeenth century. Some 200,000 tonnes were being transported annually on the canal. It has been modernised and is still in use today.

No doubt impressed by the success of the Briare Canal, Louis XIV's brother, the Duc de Orléans, commissioned Sebastien Suchet, a former student of the physicist Mariotte, to design a canal to link the Loire at Orléans to the Loing at Montargis, a distance of 74 km. Constructed between 1682 and 1692, the canal climbed 30 metres from Orléans to its summit through eleven timber locks, and descended 40 metres to the Loing through seventeen locks. Masonry locks replaced the timber locks in 1726. As the 32-km-long feeder channel to the summit had a very shallow gradient, water inflow was slow and, as with the Briare Canal, the summit section of the canal doubled as a reservoir, with the water level allowed to build up during the night for use during the day. The canal served largely to transport timber from the forest of Orléans to Paris.

Completion of the Orléans Canal placed additional traffic pressures on the Loing between Montargis and the Seine, which could be negotiated only by virtue of twenty-six flash locks built into mill weirs at various points along the river. Delays, which became intolerable, spurred the excavation between 1719 and 1724 of a lateral canal alongside the Loing with twenty-one locks in its length of 51 km.

It is appropriate, and no accident, that the greatest civil engineering achievement occurred during the reign of Louis XIV, the Sun King. If Henry IV had not been assassinated in 1610, it is certain that the Languedoc Canal (also known as the Canal du Midi) would have been undertaken during his reign, and with the redoubtable Duc de Sully in charge of public works there is surely no doubt that it would have been completed. Great civilisations throw up impressive leaders: the enlightened Henry IV paved the way for Louis XIV, and Sully for equally outstanding statesmen in Cardinal

Richelieu and Colbert. On Louis XIV's accession to the throne in 1642, the reluctance to undertake major public projects obtaining during the reign of Louis XIII disappeared, giving way to a golden age with France leading the world in civil engineering enterprise, and it has continued to be a leading player to this day.

Pierre Paul Riquet had prepared his ground well, and certainly knew of previous plans to exploit the Garonne and Aude rivers to link the Atlantic and the Mediterranean, before writing to Colbert in 1662 proposing the construction of the Langedoc Canal. Aware of the shortcomings in his own knowledge of such works, he consulted with fontainier Pierre Campas, who, with his son, also Pierre, had responsibility for maintaining local district water supplies and, in particular, knew well the streams of the Montagne Noire, from which would have to come the water to supply the many locks on the canal. In addition, he enlisted the help of a very able young engineer, François Andreossy, and tramped the countryside, making a thorough survey of the route. The analogy with another remarkable Frenchman, Ferdinand de Lesseps, 200 years later, is compelling. Also middle aged, with no engineering training (he was a retired diplomat), de Lesseps tramped the Isthmus of Suez exploring a route for the Suez Canal.

Riquet's plan appealed to the far-sighted Colbert, Louis XIV's Minister of Finance, who would have had little difficulty in persuading his monarch to accept a scheme that must have seemed God-sent to a king imbued with a vision of a glorious France. Encouraged by the King's appointment of a commission to investigate his plans, Riquet set to work during 1663 and 1664 to flesh out the bones of his proposal and submitted his final report in November 1664. The Commission accepted this report with the modification that the entire length from Toulouse to the Mediterranean should be excavated canal, instead of making navigable some minor rivers along the route and incorporating them into the scheme, as proposed by Riquet. The Commission also decided to terminate the canal not at the mouth of the Aude River, as proposed by Riquet, but at the Etang de Thau, a great sea-water lake 17.5 km long, separated from the Mediterranean by a sandbank. This required a channel to be excavated in the lake bed to link the canal to the new port of Séte constructed on the sandbank.

Recognising the importance of providing plentiful water to the summit, Colbert insisted that the main work on the canal should be preceded by a pilot channel to confirm the adequacy of the supply. Riquet undertook this at his own cost with complete success. From Toulouse to Séte the canal has a length of 240 km; it rises 63 metres from the Garonne at Toulouse through twenty-six locks over a distance of 51 km to its summit. After negotiating a relatively short summit length of 5 km, it then drops 189 metres through seventy-four locks over a distance of 184 km to the Mediterranean. As specified in the original contract plans, the channel initially had a base width of 10 metres, a water depth of 2.6 metres and side slopes of 1 vertical to 1 horizontal, but after the occurrence of several slips the side slopes were flattened to 1 vertical to 2½ horizontal. As the base width remained at 10 metres, this greatly increased the amount of excavation.

Collapse of the masonry walls of the locks early in the work led to modifications in their design, comprising a reduction in wall height by one-third, strengthened foundations and, most interestingly, curvature of the walls in plan to provide an arching action against earth pressure. This solution to countering the earth pressure gave a design

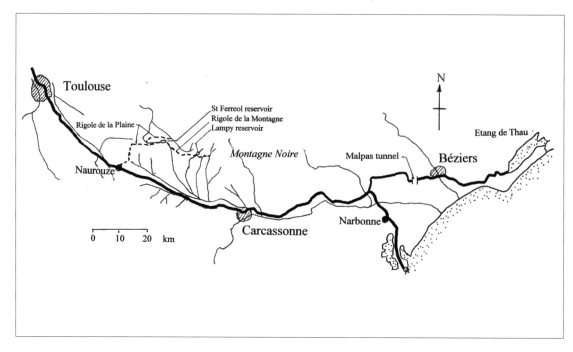

Route of the Canal du Midi. The 240-km-long canal rises 63 metres through twenty-six locks to its crest 51 km from the Garonne River, then descends 189 metres through seventy-four locks to the Mediterranean.

remarkably similar to that shown by the Italian engineer Vittorio Zonca in his *Novo Teatro di Machine et Edificii*, published in 1607. A disadvantage of the oval-shaped lock chamber was a greater loss of water. These modified locks had a length of 35 metres, an entrance width of 6 metres and an average fall of 2.5 metres. Several groups of multiple locks along the length of the canal included, near Beziers, a staircase of eight locks with a total fall of 21.5 metres.

Awarded the contract to construct the canal in October 1666, Riquet had 2,000 men on site within three months of starting work at the Toulouse end. By 1669 this had increased to 8,000 men, the canal having been divided by Riquet into twelve sections, each under the control of an Inspecteur with a gang of labourers and freemen under his command. Riquet appears to have demonstrated extraordinary organisational skills, but was not exempt from the problems that commonly beset massive projects of this type – including opposition from land owners and shortage of funds. The finance for the work came partly from local taxes in Languedoc and partly from the Royal Treasury. Riquet also committed large amounts of his own money to the project, driving himself into debt. He managed eventually to convince the local landowners of the advantages of having the canal traversing their lands, to the extent of obtaining substantial loans from them.

The canal intersected innumerable small streams, most of which were taken by culvert under the canal; three major aqueducts negotiated larger streams, two of these designed

Locks on the Canal du Midi.

by Vauban, more celebrated for his harbour and defence works. Many bridges were also built to take existing roads across the canal. Throughout the work Riquet had to answer to the Commissioners appointed by the King, and arguments he had with them over details of the route slowed the rate of progress of the project. The Commissioners deemed impossible his proposal to take the canal through a tunnel at Malpas and instructed him to take another route. Diverting their attention by sending most of the workmen to another part of the canal, he drove the 161 metre long tunnel through the hill with a small gang of hand-picked men. This tunnel also had the distinction of seeing for the first time in Europe the extensive use of gunpowder for blasting the rock.

Provision of adequate water to the summit required civil engineering works of a magnitude comparable with the canal itself. Suitable sources of water existed in the rivers of the Montagne Noire, to the north of the canal summit, but the excess water available in the winter was offset by a shortage in the summer. Riquet adopted the logical solution of creating a reservoir to store the winter surplus for use during the summer. In doing so, he had constructed the first dam in Europe to store water for a summit canal, and also the first dam in Europe to exceed 100 feet (30.5 metres) in height. He also had nearly 70 km of artificial channels, comprising the Rigole de la Montagne and the Rigole de la Plaine, excavated to collect waters from the streams of the Montagne Noir and from the St Ferreol Dam and convey them to the Naurouze summit of the Canal du Midi.

The St Ferreol Dam across the River Laudot has a maximum height of 32 metres, a base width of 137 metres and a crest length of 780 metres. Its unique construction consists of three masonry walls running the full length of the dam across the valley and firmly embedded to a depth of 3–4 metres into the rock foundation. Two of the walls,

Malpas tunnel on the Canal du Midi. Over 160 metres long, this is the oldest canal tunnel in the world.

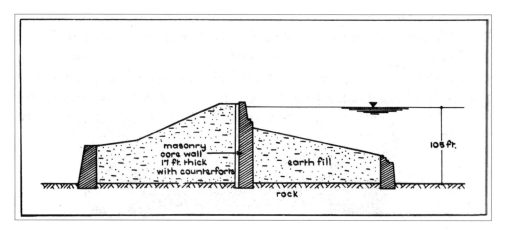

The 32 metre high St Ferreol Dam, impounding 7 million cubic metres of water, is the primary source for the Canal du Midi lock operations.

14.5 metres and 10.5 metres respectively in height, support the upstream and downstream slopes consisting of well compacted earth and stone fill, which forms the main body of the dam. The third masonry wall constitutes the core wall of the dam: it rises to the full height of the dam and its top, 5 metres wide, forms the crest of the dam. A layer of clay 2 metres thick covers the upstream slope to help reduce leakage.

Using only human labour, both men and women, to excavate, carry and place the fill, Riquet completed the dam in four years. Impounding 7 million cubic metres of water, its conception and construction constituted a remarkable achievement, and it is still in service today, one-third of a millennium after its completion, without ever having given even the slightest trouble. Towards the end of the eighteenth century, increasing traffic on the canal forced the construction of a second dam, entirely of masonry, across the Lampy River about 13 km to the east of the St Ferreol reservoir. A dam at this location had been proposed a hundred years earlier. Although impounding only about one-third the amount of water as the St Ferreol Dam, the additional capacity provided by this 16 metre high buttressed masonry dam, constructed of granite blocks, shortened from eight weeks to two weeks the canal closure period during the summer.

When Louis XIV opened the canal in 1681, Riquet had died seven months previously, but not before being ennobled by the King, becoming Baron Riquet de Bonrepos. His son Jean-Mathias saw the work through to completion. Later, it changed its name to the Canal du Midi, by which it is known today. In the field of transportation it was the first great piece of modern civil engineering and ushered in the age of major canal building

Lampy Dam, 16 metres high, is composed of granite blocks. Although recommended by the Royal Commission for construction a hundred years earlier, it was not built until 1777–81, to augment the water supply for the Canal du Midi and reduce canal closure in summer from eight weeks to two weeks.

in Europe, which in turn gave vital impetus to the industrial revolution of the eighteenth century.

Any reference to the Canal Age in Britain usually implies the period from about 1760 to 1840, embracing the period of the Industrial Revolution and commencing with the opening in 1761 of Brindley's 10½-mile-long level canal to convey coal to Manchester from the mines at Worsley owned by the Duke of Bridgewater. It not only halved the cost of coal in Manchester from 7*d* to 3½*d* per hundredweight, but ensured the industries of the city a regular supply of the vital fuel. The following decades saw a burgeoning in the construction of canals to feed the raw materials to the new industries, and to convey the finished products to the home markets or to ports for overseas distribution.

In fact, river improvements and canal construction for navigation had been actively pursued in Britain from medieval times through to the advent of the Canal Age, and indeed the first pound lock in the country was constructed in 1566. Prior to 1600, some 90 miles of river lengths made navigable by engineering works had been added to the 650 or so miles of natural river lengths navigable by river-craft. This excludes tidal lengths of river and river mouths navigable by sea-going craft such as the Avon to Bristol and the Thames to London Bridge. The next century and a half up to the start of the Canal Age saw an upsurge in activity, with the creation of a further 600 miles or so of river lengths made navigable by engineering works that included, as well as river straightening, dredging or scouring and new cuts, numerous weirs, sluices and flash locks, and at least fifty pound locks. Many of the locks, both flash locks and pound locks, served the purpose of negotiating weirs supplying water at elevated levels to power the numerous mills sited along the rivers.

Around 1280 the Countess of Devon, Isabella de Fortibus, had a weir built across the River Exe that had the effect of cutting off access to the port of Exeter from the sea. Although access was restored a decade later, her cousin, the 9th Earl of Devon, had a new weir built in 1317 that once again blocked access of shipping to Exeter. He also had a quay built at Topsham, nearly four miles downstream from Exeter, and profited greatly by imposing tolls on transportation of goods overland from this quay into Exeter. Despite many petitions from the city over the next two hundred years and more, it wasn't until 1550 that King Edward VI granted permission for the waterway to be reopened, an empty gesture as the river channel had now silted up. This led the frustrated Exeter traders to commission a canal 1¾ miles long to be built, linking the city with the river downstream of the weirs, employing John Trew of Glamorgan to carry out the work, which he completed in three years from 1564 to 1567.

The canal excavation, partly embanked, had a width of 5 metres and a depth of nearly 1 metre, sufficient for boats of up to 16 tons burden. In creating this canal, Trew built the first known pound locks in England, numbering three in this short length of canal. All were turf-sided, with mitre gates at one end and single guillotine gates at the other end, the lower two locks being 57.5 metres in length and 7 metres wide, with 1.7 metres depth of water, able to accommodate several vessels at a time and to act as passing places. The upper lock, even larger, served as a basin or dock. The quayside in Exeter consisted of a 46 metre long masonry wall. Over a hundred years after its completion, the canal was extended downstream to Topsham, and deepening and widening early in the eighteenth century, together with a reduction to a single lock, allowed access to ocean-going ships. A timber weir across the river built by Trew provided adequate water to operate the locks and maintain water level at the Exeter quay.

Weirs built by mill owners were commonly 1.5–3 metres in height and passage of river craft past them meant unloading on one side of the weir and reloading on the other, or the incorporation of a lock or 'turnpike', usually a timber flash-lock, to enable the craft to negotiate the structure. With the raising of the turnpike, the craft had to wait for the rush of water to abate before attempting the upwards or downwards passage. In some cases their passage could be held up for days or even weeks if a mill owner chose to delay opening the turnpike for any reason, in addition to which the boat owners or bargemasters could be subjected to excessive charges by the mill owners.

Waltham Abbey, a settlement about eleven miles to the north of central London, received water to power a mill from the River Lea by means of a millstream taking off from the river nearly one mile upstream, a turnpike flashlock diverting the flow of water at this point. In 1574 the Commissioners appointed to oversee the improvement of the River Lea navigation ordered that boats in future be directed into the millstream to avoid having to operate the little-used flashlock and having to scour the river bed from time to time. This necessitated the construction of a new 180 metre long cut linking the millstream at a point near the mill to the main river nearly a mile downstream from the turnpike and incorporating into the cut a pound lock. This became the first known walled lock to be built in England and the first to have mitre gates at each end. The timber walls and mitre gates enclosed a chamber some 18 metres long, 6.7 metres wide and 3 metres deep, built at a cost of £272. After three years it had to be reconstructed on masonry foundations to overcome settlement problems, an operation carried out under the supervision of Mr Trewe, almost certainly the Exeter Mr Trew. The Commissioners also ordered the replacement of the flashlock by a loweshare, a form of submerged weir, to raise the water level.

The lock attracted the notice of the poet William Vallans in the 1580s, who imagined what two swans journeying down the River Lea would have marvelled at:

> But newly made, a waterwourke: the locke
> Through which the boates of Ware doe pass with malt.
> This locke contains two double doores of wood,
> Within the same a Cesterne all of Plancke,
> Which only fils when boates come here to passé
> By opening of these mightie dores with sleight,
> And strange devise, but now decayed sore.

Its construction caused resentment among local fishermen and land-owners, the latter because it enabled London brewers to now obtain their grain quickly and cheaply from the Midlands. The most violent opposition to the scheme, however, came from the badgers, who carted the local agricultural produce to London by road and saw their livelihood being taken from them. They damaged works along the river, forced barges to stop using it and set fire to the lock, though with no great effect. Ultimately the mill owner, Edward Denny, defying the Commissioners, had the lock demolished, whereupon the bargemen attempted to pull up the loweshare, allowing them to proceed down the old river route, which again angered the badgers, leading to further violence, damage to and sinking of barges and injuries to men on both sides. The bargemen appealed to the authorities and won their case, a decision immediately exploited by Edward Denny, who had the turnpike flashlock rebuilt and collected a toll of 5s from each passing barge.

Extensive engineering works carried out in the seventeenth and early eighteenth centuries put England at the forefront of countries in Europe utilising their rivers for navigation purposes. A 15-mile stretch of the Thames from Burcot to Oxford, with many shoals, was made navigable for large barges in 1638 by the retention of one existing flash lock and the construction of three pound locks 24 metres long and 5.8 metres wide, with masonry walls and mitre gates. The Wey navigation, completed in a period of two years between 1651 and 1653, comprised perhaps the most impressive river engineering works of the seventeenth century, its total length of 15 miles consisting of 9 miles of cut, ten pound locks, four new weirs, twelve bridges and a wharf at Guildford at its southernmost end. Lock rises ranged from 1.2 metres to 2.4 metres and accounted for 18 metres of the total rise of 22 metres to Guildford from the tributary Wey's inflow into the Thames at the northern end of the works. Designed by Sir Richard Weston, the total cost exceeded £15,000 and employed up to 200 men, his son taking over direction of the works after the death of Sir Richard in 1652. Engineering works, completed in 1723 on another tributary of the Thames, the Kennet, required 11 miles of cuts, with twenty pound locks along its 18½ mile length to bypass mills sited on the river. At least two of the locks were of brick construction and the remainder were timber-framed or turf-sided, all of them able to accommodate fully laden 100 tonne barges bound for London from Newbury. Falls ranged from 1.5 metres to 2.8 metres. Other new works included wharves at Newbury, Aldermaston and Reading, a barge basin at Newbury, several bridges spanning the cuts and some dredging and some lengths of embanking. In 1814 the Scottish-born civil engineer John Rennie completed the construction of one of the great achievements of the Canal Age in England, the 57-mile-long east–west Kennet and Avon canal, linking the Kennet river navigation at Newbury to the Bristol Avon river navigation scheme at Bath, completed earlier in 1727. Unlike the Kennet navigation, the pound locks on the 11½-mile-long Avon river, six in number and masonry-walled, 30 metres long and 5½ metres wide, were located in short cuts immediately adjacent to the mill weirs. Five of the locks had drops of 1–1.5 metres, but one lock accommodating two weirs had a drop of about 2.8 metres.

5

Traversing the Land

The rapid spread of Islam created the need, and in some cases simply the urge, to travel. Trade and communication, not only between Islamic countries, but also with non-Islamic states such as the Byzantine Empire, required the movement of people and goods. The latter included a wide range of foodstuffs, woollens, cotton and silk textiles, carpets and tapestries, dyestuffs, perfumes, manufactured metal goods and timber products. Scholars, philosophers and scientists travelled to acquire knowledge or exchange ideas. The Hajj, the pilgrimage to Mecca required of every Muslim able to do so once in his or her life, generated much of the travel and contributed to the widespread cultural unity within Islam, as did the adoption of one religion and the use of one language. The widespread usage of the Arabic language effected a unifying factor which neither the Greeks nor the Romans were able to achieve with their languages in the daily lives of the peoples within their influence or under their control. An efficient postal service, established primarily for the transmission of official letters, also served to keep the central government informed on concerns such as land taxation and the work of officials in the provinces, as well as fiscal, economic and political matters. The postal service also found other uses, such as the transportation of ice from Syria to Egypt for use in cooling drinks. Some use could be made of surviving lengths of the Roman road system but except for main streets in cities, they did not need paved roads as the hard surfaces were neither necessary nor desirable for the pack animals – camels, donkeys, mules and horses – almost universally used for transportation and portage. The postal service relied almost exclusively on mules and camels, with staging posts set up 12 km apart in Iran and 24 km in western provinces to facilitate travel. Enterprising Muslim traders took their goods overland deep into equatorial Africa, into southern Russia and their enormous camel trains even made it as far as the borders of India and China.

As the centralised authority of Rome disintegrated in western and northern Europe, so did the need for roads, and indeed it did not serve the interests of the various dispersed centres of power to maintain efficient road systems that could assist the rapid movement of enemy forces. Thus, the great Roman road system fell into neglect. It says much for the high quality of the original construction that, despite little or no maintenance, or indeed a real understanding of the principles of good road construction, appreciable lengths of the gradually deteriorating Roman road system remained in use throughout the medieval period in Europe. Some roads are still in use today.

As in Islamic countries, wheeled traffic other than handcarts was very little used for land transportation of goods in Europe, reliance being placed mainly on human portage and pack animals, particularly horses, which moved with greater facility on gravelled or compacted earth surfaces than on hard flagged stone pavements. The situation became self-perpetuating, as the continuing deterioration of the roads encouraged the use of pack animals, and any incentive there might have been to improve the roads disappeared. The light handcarts in common use did not need a high-grade road surface. Cumbersome oxen or horse drawn carts, used for conveyance of bulky and heavy goods such as grain, thanks to their massive construction and slow, labouring movements, had the ability to absorb heavy jolting and shocks from uneven road surfaces.

Not only were the Roman roads allowed to decay by natural processes, they often suffered positive deprivation at the hands of farmers and villagers, who were not above plundering them for building stone. Even poor quality roads did not escape these quarrying activities, sometimes with tragic consequences. In 1499 an Aylesbury miller in England, Richard Boose, sent a couple of servants to dig from the middle of a nearby road 'ramming clay' which he needed to effect repairs to the mill. Undertaking their task with commendable vigour, they left behind a hole some 3 metres wide and 2.5 metres in length and depth, its presence unfortunately disguised by the muddy water that quickly filled it, so that it looked like just another inconvenient shallow puddle. A glover from Leighton Buzzard, his horse laden with panniers of his wares for the Christmas market in Aylesbury, happened along shortly after, having no doubt already splashed his way through many puddles. Man and horse disappeared into the pit and drowned. Charged with the man's death, the miller pleaded no malicious intent and that this was the only source of the kind of clay he needed. He gained acquittal on these grounds.

Although it was considered a pious duty of the Church to show charitable concern towards the traveller, which included road maintenance, it often appears to have fallen down on the job. So, too, did the landed proprietors, who had nominal responsibility, at least, for roads adjoining their holdings; they simply transferred their responsibilities to their tenants, who had even less motivation to do work from which they derived little or no benefit. Many of them had no concept of a road system or road construction and throughout the whole of their lives never travelled beyond the nearest market town, to and from which they would transport their produce for sale, or occasional purchases, by their own human portage supplemented, at best, by a handcart or pack animal. Their pace of life did not demand rapid travel and they became accustomed to finding their way around potholes or puddles (or occasionally into them, as in the case of the unfortunate glover), boggy ground or even, in some instances, trees growing in the middle of the roadway. Similarly, these hazards did not present a problem to sheep, cattle or goats being driven to market and their droppings became simply another hazard for following travellers to avoid if they could.

Europe was much more heavily forested in medieval times than today, so that many of the roads comprised little more than cleared ways through the forest. The road usages called not for a high quality durable surface supported by a carefully built-up and compacted pavement structure, but for a cleared way of sufficient width to allow quagmires, potholes and other obstacles to be avoided and to enable the road user to keep adequate distance between himself and bushes, hedges and other cover fringing the road where robbers could well be lurking. In Germany the extent and denseness of the forests made this form of roadside clearance impractical, and furthermore the ubiquity of the timber led to the unsuitable use of its trunks and branches for repairing roads in place of stones or gravel. The Marburg to Frankfurt road, in 1571, developed a deep hole that claimed three wine carts and the life of a labourer. It took 500 bundles of brushwood weighted with stones to fill it.

Notwithstanding the generally poor condition of the roads, there were important interests that made sure some of the major road links stayed in passable state of repair. These included the interests of the Church in maintaining communications between the great ecclesiastical houses, which in Britain ensured attention to selected parts of the road system up to the time of the dissolution of the monastic houses by Henry VIII. The occasional new length of pavement construction or reconstruction usually consisted of no more than a layer of gravel or broken stone over a bed of sand. Within the environs or approaches to the monasteries or towns, a more durable pavement might be constructed of stone blocks set in mortar on a bed of sand.

Up to about the middle of the sixteenth century, travellers in Europe, with few exceptions, went on foot, or on horseback if sufficiently wealthy or well-connected. The use of heavy, wheeled carriages, in fact little more than four-wheeled wagons with lavish added appointments, was limited to noble or royal ladies with their retinues, although more often they elected to be carried by litter or to ride their own palfreys, sidesaddle. By the end of the sixteenth century, however, coach travel had become much more common, prompting John Stow in his *Survey of London* to observe:

> Of old time coaches were not known in this island, but chariots or whirlicotes, then so called, and they only used of princes or great estates, such as had footmen about them ... but now of late years the use of coaches, brought out of Germany, is taken up and made so common as there is neither distinction of time nor difference of persons observed, for the world runs on wheels with many whose parents were glad to go on foot.

This proliferation of coaches with their iron-shod wheels, and often unsprung suspension, contributed to a rapid deterioration of road surfaces. In 1621, in the time of James I of England, concern about the increased use of wheeled traffic led Parliament to pass an Act limiting the vehicles which could use the roads to two-wheeled carriages only, and limiting to one ton the maximum load any vehicle could carry. This clearly offered a cheaper solution than improving the roads.

The most frequent users of the medieval roads included, notably, the European kings and their courts. In times of civil war the roads served both the king and those opposed to him; in times of peace they enabled kings to pursue personal interests and the business of state. Hawking counted as one of the great passions of medieval kings and they didn't like to be stopped from following their birds by poor road or bridge conditions. Records survive from 1373 of Edward III of England ordering the Sheriff of Oxfordshire to have all bridges repaired and fords marked with stakes for the crossing of the King 'with his falcons' during the approaching winter season. It was Edward III, too, who in 1373 ordered the paving of the 'highroad' running from Temple Bar to Westminster. Although a paved street, the King observed it to be 'so full of holes and bogs … and the pavement so damaged and broken' that it had become dangerous for men and carts. Land-owners adjoining the road were ordered to make footpaths 7 feet wide and a tax was levied on all merchandise proceeding to Westminster to cover the expense of paving the middle portion of the road. Presumably encouraged by the income derived from this tax, the City authorities, three years later in the thirteenth year of Edward's reign, imposed a tax on all vehicles and laden beasts entering or leaving the city. The charge was a penny per cart or a farthing per horse, with reductions for multiple passages; a cart bringing sand, gravel or clay paid three pence per week. 'No charge was made for the carts and horses of Great People.'

The poor condition of London streets made travel by wheeled vehicles uncomfortable at best for passengers and even dangerous, and to be avoided as far as possible. Where possible, many people took advantage of the Thames to reach their destination. The nobility and the wealthy also used sedan chairs for conveyance. Inhabitants of some towns and cities in England and Scotland fared rather better. One traveller in 1640 described streets in Leicester, which had roadworks done by contract, as 'so well paved, and kept so clean from dunghills, filth or soil, that in the wettest and foulest weather a man may go all over the town in a pair of slippers and never wet his feet'. Another worthy, Sir William Brereton, described the High Street of Edinburgh, now known as the Royal Mile as 'the best paved street with boulder stones, which are very great ones, that I have seen. The channels are very conveniently contrived on both sides of the street, so as there is none in the middle; but it is the broadest, largest and fairest pavement, and that entire, to go, ride, or draw upon'. Some wealthy citizens even bequeathed money to be used for the benefit of the community, such as Robert Hare, who left £600 to Trinity College for augmentation of a fund for repairing the highways in and about Cambridge.

In times of peace, active kings and their courts travelled frequently across the country conducting State business, administering justice and generally investigating conditions in the realm. The various nobles, knights, officials and accompanying ladies and retainers practically constituted a small army travelling through the countryside. With the richly caparisoned horses of the nobles and the varying and colourful accoutrements of the travellers, it would have been an impressive site. Noble ladies of the time often rode astride their horses, a mode, as shown in the fifteenth-century Ellesmere manuscript, favoured by Chaucer's Wife of Bath; but some, such as Chaucer's Prioress in the Ellesmere manuscript, chose to ride side-saddle. Occasionally a noble lady might elect to ride with her attendants in a cumbersome, but luxuriously appointed, carriage. Drawn by three or four horses, the sheer size and weight of these carriages would have absorbed some of the thumps and bumps meted out by the uneven road surface; with the welfare of the ladies further ensured by their being able to recline elegantly in comfortable chairs generously furnished with deep, beautifully embroidered cushions. Internally, such carriages were usually bedecked with dazzling tapestries and the window openings were hung with silk curtains.

In medieval times, kings and nobles rode on horseback, as did some of their ladies, but others of the ladies chose to travel by horse litter or in lavishly accoutred, but cumbersome, carriages such as this with their retinue. They were usually drawn by three or four horses, one behind the other, with a postillion mounted on one of them.

The progress of this brilliant cortege would have been viewed with mixed feelings by the villagers and townspeople. Marshals sent on ahead appropriated corn, hay and oats for the animals, and food and drink for the travellers, with at best inadequate compensation and at worst no compensation at all. The commandeering of accommodation for the travellers often meant the temporary eviction of existing occupants. And the rich, noble families did not escape lightly. They had to accommodate and entertain the king and his immediate entourage in a matter befitting their status.

Town officials, too, must have viewed with some trepidation the pending arrival of the king and his officials, who would have been quick enough to pinpoint examples of dereliction of public responsibilities. In Cambridge in 1335, the King ordered the Mayor and bailiffs to repair the paving of streets and lanes and to compel owners to pave the streets in front of their tenements. No doubt Edward III felt well justified in making this order, as Cambridge had been granted the right by Edward I, in 1289, to levy a 'pavage' on many classes of goods using the streets, for their maintenance and repair. Several other towns in England acquired similar rights.

Paving became a highly respected, expert job commanding a reasonable stipend fixed by ordinance. In busy streets, markets and squares, the paving usually consisted of cobbles embedded in sand and placed on the existing surface after filling potholes. This led to a gradual raising of street level. Gravel and sand compacted with hand-rammers were used for paving less important streets and lanes.

Although much of the trade generated by the Hanseatic League went by sea or by boat along inland rivers or canals, overland routes also assumed considerable importance, notably in Germany and France. The success of the great commercial medieval Fairs, where most of the merchants conducted their business, depended upon the state of the roads. There exist many examples of towns, such as Nüremberg in Germany, which owe their medieval origins to their locations as key trading centres with road links to many parts of the country. It was in the interests of these great trading towns in Europe to keep

A sixteenth-century illustration depicting the construction of cobbled paving.

their approach roads in good condition. In some instances they took responsibility for the improvement of roads many miles away, but which gave vital access from another major town. Several miles of the Ghent road out of Paris were repaired and improved in the fourteenth century by the town of Ghent. Sufficient commercial pressures could overcome obstacles, no matter how challenging, as instanced by the opening up in the fourteenth century of a road capable of taking wheeled vehicles through the high Septimer Pass in the Alps.

Goods conveyed by road were multitudinous and varied: coal, timber, beer, wines, grain, sheep, cattle, pigs, horses, butter and cheese, fish, furs and skins, silks and woollen cloths, manufactured goods such as pots, millstones and many others. But, as on the rivers and canals, innumerable tolls, particularly in Germany and to a lesser extent France, gradually choked off much of the trade and had the effect of encouraging local self-sufficiency. The Landfriedgesetz or Peace Law introduced by the German Emperor Frederick II in 1235 abolished the mandatory use of specific roads, but this apparently enlightened decree exercised little authority in specific areas where local laws confined vehicles to travel on certain roads only, and threatened fines for any attempts to avoid toll-collection points. Other constraints imposed on merchants included requiring them to first offer their goods for sale in their local town and, once on the highway, to display them for sale for three days in any town they passed through that had a recognised market. Anyone not doing so could face a substantial fine. This threat may have seemed relatively insignificant compared to the hazards of highway robbery on the roads by bandits, sometimes operating in organised gangs, or less organised army deserters or out-of-work labourers. Noblemen, too, were not above committing such robberies, either to ease their impoverishment or simply as a matter of sport.

A post and passenger coach service established early in the sixteenth century linked a number of the major cities in Europe, but does not seem to have led to any marked immediate improvements to the roads. The system set up in 1504 by Franz von Taxis, originally from Bergamo, provided a postal service linking the court in Vienna of Maximilian I, Holy Roman Emperor and German King, with the Low Countries and the courts of France and Spain. By 1516 the fortnightly service had a network extending to several other major towns including Brussels, Naples and Rome, Nüremberg, Frankfurt and Hamburg. Originally only court officials or those on imperial business could travel as passengers on the coaches, but within a few years this privilege was extended to a wider range of persons, and weekly services were instituted.

If the amount of criticism and abuse directed against Europe's roads in the sixteenth to eighteenth centuries is any guide, road conditions had deteriorated to an even worse state than in the medieval period. In fact, this may have been more a reflection on people's expectations and, perhaps more importantly, the increased use of heavy wheeled traffic, particularly for passenger conveyance and postal services. The gradually increasing pace of life made delays more frustrating. Even so, deterioration of the roads may have occurred and, in Britain at least, this can be blamed partly on the dissolution of the monasteries by Henry VIII in 1536-9. The new owners who purchased the landed properties of the monasteries from the Crown did not share the view of the Church that the upkeep of the roads constituted a pious duty.

A limited amount of information and tables of English roads became available in the sixteenth century, and 1579 saw publication by Christopher Saxton of the first book containing maps of England and Wales, but not showing the roads themselves. In 1625 Norden published *An Intended Guide for English Travellers*, with tables for each county showing the distance between towns. In 1675 John Ogilvy published a large folio containing strip maps for the roads throughout England, with descriptions of road conditions and landmarks.

King James I, concerned about the increased use of wheeled traffic on English and Scottish roads, passed a decree limiting the vehicles that could use the roads to two-wheeled carriages only. This clearly offered a cheaper solution than improving the roads. Charles II modified the restrictions in 1661, but stagecoaches had already been running between London and Edinburgh since 1658. It was about this time, too, that the first turnpikes came into existence in Britain, introducing the novel idea for the British of charging road users for the use of the road. The name Turnpike derived from the swinging toll barriers made of sharpened poles.

The first Turnpike Act to be passed in England, in 1663, was 'For Repairing the Highways within the Counties of Hertford, Cambridge and Huntingdon'. It met with some strong opposition and the proposal to erect gates at Stilton in Huntingdon stirred protests violent enough to ensure its being shelved. The Caxton Gate in Cambridge proved ineffective as it could easily be avoided. By 1750 no less than 169 Turnpike Acts had been enacted and this increased rapidly to 1,100, covering 20,000 miles of road by 1840, but much of the population did not acquiesce readily to these Acts, as evidenced by the riots around the middle of the eighteenth century which resulted in the destruction of many turnpikes and toll-houses. Blood was shed as troops attempted to quell these violent outbursts.

Nevertheless, the establishment of the turnpike system in England did lead to an improvement in road construction and maintenance, due in no small measure to the appointment of full-time surveyors to assume responsibility for their upkeep. Britain was ready for a breakthrough in road construction and, as so often in these circumstances, it threw up the men capable of designing and building better roads. 'Blind Jack' Metcalfe, Robert Macadam and Thomas Telford became household names in the second half of the eighteenth century and the first half of the nineteenth century. Some of the principles put into effect by these great road builders had been anticipated by Robert Phillips, who read his dissertation concerning *The State of Highroads in England* before the Royal Society in 1737. He recognised, for example, that a layer of gravel resting on a well drained dry 'sole' would be beaten by the traffic into a firm road surface, which described, in effect, the Macadam Method.

In France some road building activity took place in the sixteenth and seventeenth centuries that eventually led to an upsurge in road construction, firstly in France in the eighteenth century and subsequently in other parts of Europe. A road connecting Paris to Orléans built in 1556 had a full width of 16 metres, with the middle strip of 4.5 metres paved. The enlightened plan of Henry IV and his minister the Duc de Sully to develop a system of inland waterways in France extended also to road construction. In 1597 Sully initiated work on a national system of roads, paved with broken stone, but, as with the waterways, Henry's successor Louis XIII stopped the work. The publication of a book, *Histoire des Grand Chemins de l'Empire Romain*, by a French lawyer, N. Bergier, based on road remains he discovered near Rheims revived an interest in Roman road construction methods. It had little immediate effect, but did eventually influence road construction in France. As with the waterways, it was again Colbert, Louis XIV's Controller General of Finances, who saw the wisdom of Sully's intended network of roads, and in 1661 he endeavoured to introduce a programme of road construction and maintenance, but apart from a few roads around Paris little progress was actually made until around 1700.

Credit must go to Hubert Gautier, an outstanding French engineer, for establishing France as the leading country in the world for road construction in the eighteenth century. In 1693 he published an article on road construction entitled 'Traite de la Construction des Chemins' which, together with the establishment in 1716 of the Corps d'Ingenieurs des Ponts et Chaussees, gave a much needed stimulus to road building. Gautier was appointed Inspector of Ponts et Chaussées at the time of establishment of the Corps and remained in the post until his death in 1731 at the age of 71. He advocated a modified Roman design for pavement construction on a prepared flat subgrade, typically with a width of 5.5 metres and a depth at crown of 450 mm, sloping off to 300 mm at the edges. Flat stones in two or more layers laid at the base of the excavation supported two more layers of smaller broken stones. In order to pay for this expensive form of construction, a poll tax was imposed on the owners of properties bordering the roads. It also made great demands on labour, leading to the use of impressed or corvée labour, women as well as men. This helped to foment, in some measure, the unrest that finally brought about the Revolution in France, although compulsory road labour had already been abolished in 1764, twenty-five years before the storming of the Bastille.

Modern road construction owes its origins to the principles established by Tresaguet and his British counterparts and also, in no small measure, to the establishment in France in 1747 of the Ecole des Ponts et Chaussées. French engineer Pierre Tresaguet introduced a cambered subgrade around 1764 to shed water penetrating through the pavement into deep side-ditches, rather than allowing it to soak into, and thus soften, the subgrade soil. This greatly reduced the required pavement thickness and cut significantly the cost of road construction. On the subgrade he laid a foundation of stone blocks on edge and on this placed a uniform thickness of broken stone. The top layer, also of uniform thickness, giving a cambered surface the same as that of the subgrade, consisted of a layer of small broken stones. This form of construction found widespread use in France and Britain, in the latter case promoted by Thomas Telford among others, until superseded by the cheaper Macadam method of placing a compacted layer of broken stone immediately on the cambered subgrade, surmounted by a carpet of smaller stone in which the fine material produced under the action of traffic occupied the interstices and bound the surface.

6

A Span of Bridges

Bridges surviving today from the pre-industrial age are overwhelmingly masonry arch structures, whereas many of the bridges built during these times would have been of timber. Unless constantly maintained or rebuilt, these will have long since disappeared through the natural processes of decay or destruction by flood, fire or military action. Indeed, many of the masonry bridges seen today followed a succession of timber structures at the same sites.

Timber Bridges

Timber has excellent qualities as a building material, with the ability to resist tension, compression and bending forces; but its strength limitations restrict simple timber beams to spans of about 6 metres. Above this, the self-weight of the beam, together with the imposed load, causes sagging or even breaking of the beam. As timber offered advantages in handling and speed of construction, and was sometimes the only suitable material readily available, there must have been a constant challenge to use it for spans in excess of 6 metres and avoid a multiplicity of mid-stream supports. Possible forms to achieve this included the arch, the cantilever and the truss.

In his famous lecture to Ludovico Sforza while seeking employment in Milan, Leonardo da Vinci claimed among other things to 'have a sort of extremely light and strong bridges, adapted to be most easily carried, and with them you may pursue, and at any time flee from the enemy; and others, secure and indestructible by fire and battle, easy and convenient to lift and place'. He clearly had timber structures in mind.

In his sketches, Leonardo shows several examples of timber bridge structures, some of a temporary form, offering quick erection, and thus presumably for military purposes, and others of a more permanent type. His temporary bridges embraced simple arch and trestle forms, including an example of a trestle bridge with built-in jacks to allow the heights of the longitudinal stringers to be adjusted at one end to accommodate different bank heights. Triangulated trusses are illustrated in his sketches for more permanent bridges, one of the examples having a double deck system. His sketches also include a swing bridge consisting of a combined truss and arch pivoted about one end. It is a matter of conjecture whether or not any of these structures were ever built.

It can be safely assumed that timber cantilever bridges found common use in Europe, in medieval and Renaissance times, particularly in the form of horizontal timber beams, perhaps up to 15 metres long, with inclined struts projecting out from the piers to support the beams at or near their one-third points. Parsons illustrates such a structure designed in the late sixteenth century to be carried on stone piers across the Bisenzio River in Italy. At about the same time, Andrea Palladio employed this type of structure, also known as a straining beam truss, for a bridge over the river Brenta near Bassano in Italy. For spans of only about 15 metres this form of construction sufficed, but faced later with a significantly longer span, Palladio met the challenge by applying the true truss form.

Although Palladio's fame in the architectural world rests predominantly on his revival of Greek and Roman classicism in masonry structures, his contribution to civil engineering lies in the rather different and almost conflicting area of promoting the use of timber truss for bridge construction. He did not invent the timber truss, and indeed it will never be known who built the first triangulated truss or for what purpose. Certainly before Palladio's time timber trusses had been introduced into roof constructions and had also been used for centring in the construction of masonry voussoir arches. Indeed, Palladio himself made no claim to have invented the truss form, but merely to have introduced it into Italy. He states in his *Four Books of Architecture* (published in 1570) that he learnt of the form from a friend who had seen bridges of this type in

Palladio's bridge over the Brenta River near Bassano, featuring cantilevered beams, was reconstructed faithfully in its original form after destruction in the Second World War.

Germany. His claim to invention is confined to the specific bridge trusses shown in his volumes.

Palladio describes a number of truss configurations in his *Four Books*, of which only one appears to have been constructed, over the Cismone River. He says of this work:

> The Cismone is a river, which falling from the mountains that divide Italy and Germany, runs into the Brenta, a little above Bassano. And because it is very rapid, and that by it the mountaineers send great quantities of timber down, a resolution was taken to make a bridge there, without fixing any posts in the water, as the beams that were fixed there were shaken and carried away by the violence of the current, and by the shock of stones and trees that by it are continuously carried down: therefore Count Giacomo Angaro, who owns the bridge, was under the necessity of renewing it every year.
>
> The invention of the bridge is, in my opinion, very worthy of attention, as it may serve upon all occasions, in which the said difficulties will occur; and because that bridges thus made, are strong, beautiful and commodious: strong because all their parts mutually support each other; beautiful, because the textures of the timbers is very agreeable, and commodious, being even in the same line with the remaining part of the street. The river where this bridge was ordered, is one hundred feet wide.

Simplicity of form and absence of redundancy are striking features of Palladio's truss designs, which remained unequalled in clarity until the great period of timber bridge building in North America in the first half of the nineteenth century. Palladio's concept of the manner in which the different truss members act is clearly shown in the jointing, the iron straps tying the vertical members to the lower chord indicating their role as tension members. The notched connections between the diagonals and vertical members allow the latter to take tension, but permit the diagonals to take compression only. Where they are envisaged to act as beams, chord members are taken continuously through the joints.

Palladio had the truss members of the Cismone bridge cut to give a slight upward camber to the centre, in side elevation, imparting a feeling of strength to the 100 pieds (35 metres) span and avoiding the possible illusion of sagging in the middle of such a long span. Having had this bridge proved successfully in service, he went on to design more ambitious trusses which, however, did not see service. These included a modified form of the Cismone truss with cross-bracing in the panels, and a truss with triangular end support panels and six internal rectangular panels with parallel top and bottom chords. The joining arrangement had the same form as the Cismone truss and Parsons criticises the design on the basis that partial loading of the structure would have reversed the deformations of the panels, putting tension into the single diagonal members that the jointing system would not have permitted them to carry. He stresses the need for second diagonals in the central two panels (as in the modified Cismone truss) and possibly also in the panels adjacent to these. Another criticism directed at this truss by Parsons concerned the provision of strengthening pieces attached to the bottom chords at the ends, whereas the greatest stresses in the bottom chord occur at mid-span. Palladio's remaining design was a braced arch, with the chords forming concentric circles and

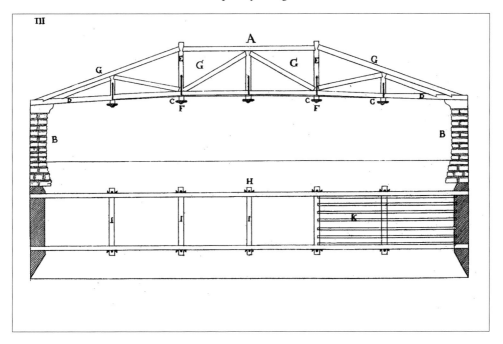

Palladio's truss bridge over the Cismone River.

the posts lying on radial lines. A double system of bracing extended throughout the structure, including the end support panels.

Blessed with extensive forests and driven by the desire to open up the country to settlement, travel and trade, the Americans in the late eighteenth and first half of the nineteenth century built covered timber bridges, beloved of poets and artists, some appoaching 100 metres in span, using a variety of truss designs for the most part, but in some cases trussed arches or with arch elements incorporated into the trusses. The covering of roof and side walls served not to shield from the elements those using the bridge, but to protect the timbers and particularly the joints of the bridges and thus prolong their life.

Covered timber bridges in Europe, predating those in America, may have been the inspiration for their widespread adoption in the New World. Although there exists no positive evidence for this, some of their designers would certainly have known of the European examples and indeed some, such as Louis Wernwag (1770–1843), were European by birth and he may well have been familiar with this type of bridge when he left his German homeland at the age of seventeen. His Colossus Bridge at Philadelphia, a combined arch and truss structure completed in 1812, boasted a record span of 104 metres.

Covered timber bridges do not appear to have been built in great numbers in Europe, and the known examples were relatively unsophisticated compared to those in the New World. First built in 1333, the present Chapel Bridge in Lucerne, much restored and rebuilt and much photographed, retains the original crudely triangulated truss form and appearance and moderate spans resting on piled foundations. The 200 metre long bridge

The covered single-span Holzbrüke over the Limmat River at Turgi in Switzerland, dating from early in the nineteenth century, features the Grubenmann principles of diagonals of varying lengths springing from the abutments with vertical queen posts and horizontal members.

angles across the Reuss River to avoid a stone water tower, and is adorned with early sixteenth-century paintings chronicling the city's history.

A number of covered bridges were built in Switzerland in the eighteenth century by two village carpenter brothers from Teufen, Johannes and Hans Ulrich Grubenmann, most notably the crossing of the Rhine at Schaffenhausen in two spans in 1757 by Hans Ulrich; a single span of 67 metres at Reichenau by Johannes; and the crossing of the Limmat River at Wettingen, a combined effort by both brothers. Unfortunately, none of these survives today, having been burnt by the French in 1799 after their defeat by the Austrians.

Defying analysis, the Schaffenhausen bridge consisted of diagonals of varying lengths springing from the abutments, together with queen posts and horizontal (or nearly so) members giving a combination of truss and arching action, the latter spanning the whole two spans of the bridge. Other diagonals radiated from the central support. As Hans Ulrich's bridge was replacing an earlier structure demolished by floods in 1754, he sensibly proposed a single span of 120 metres to cross the river, but the City Council, apparently not convinced that a simple village carpenter could accomplish this, insisted that the structure should rest on an existing intermediate pier left behind after the earlier bridge had been swept away. The story goes that Hans Ulrich removed the central support at the opening ceremony to demonstrate that it could stand as a single span. The roof structure of the gallery topping the bridge again defies analysis, consisting of a complexity of, and inefficient use of, timber members. Altogether, 400 fir trees are reported to have been used in constructing the bridge.

Johannes built a similar bridge to Schaffenhausen at Reichenau with a single 73 metre span, but when they worked together on the 60 metre span Wittingen bridge, the two brothers designed it as a true non-truss timber arch, perhaps the first ever of this type. The laminated arch ribs, about 2 metres deep, consisted of seven layers of staggered, butt-jointed oak planks tightly bound together at 2.5 metre intervals by wrought-iron straps.

The Grubenmann bridges attracted considerable attention from and visitations by engineers, architects and other interested travellers, and some may well have felt the same as the American poet Daniel L. Cady on first experiencing the Old Toll-Bridge at Springfield, Mass., one verse of which will suffice here to express his sense of awe:

You enter and the wonder grows –
The double arches spring,
And beams and braces fill the air
Like birds upon the wing;
The sky seem full of sticks and studs,
Some upright and some flat –
By George! I wish I had the sense
To build a bridge like that.

Masonry Bridges

Many Roman voussoir arch masonry bridges are still in existence today, attesting to the skill of their builders and the durability of their building material. Although adhering rigidly to the semi-circular arch form, they produced structures both functional and pleasing in appearance: the Alcantara Bridge over the Tagus River in Spain is surely one of the most daring and stunningly beautiful bridges ever built. The first-century Roman architect/engineer Vitruvius proposed three requirements in the construction of public buildings: durability, convenience and beauty. He could just as easily have proposed these as the requirements for the construction of bridges and judging by the examples seen today, the builders achieved them with something to spare. Freed from the constraints imposed by the semi-circular arch, the bridge builders of the pre-industrial age might be expected to have produced bridges at least the equal of, and even superior to, those of their predecessors, and in a few notable examples they certainly did so.

In AD 260 the Sassanian King Shapur defeated the army of the Eastern Roman Emperor Valerian and put the Roman prisoners of war to good use building a combined dam and bridge spanning the Karun River at Shustar, a town some 50 km east of Susa in Iran. The Sassanians, a Persian dynasty, thus acquired some of the technology such as adopting piers with a rubble core faced with masonry, arched openings in spandrels to increase flood flow capacity and the construction of coffer dams to enable pier foundations to be constructed in the dry. However, they eschewed the semi-circular arch in favour of the pointed arch and occasionally elliptical arches.

Muslim bridge builders took over where the Sassanians they had subdued left off, in some cases restoring or rebuilding ruined Sassanid bridges. An account by the thirteenth-century writer al Qazwini describes the construction of the great single-arch span at Idhaj in Khuzistan and at the same time contains a reminder of that ever present hazard threatening the life of all ancient masonry structures in Iran and in many Mediterranean countries – earthquakes:

> Idady is a country lying between Ispahan and Khuzistan, subject to earthquakes. The greatest wonder there is the bridge Harah Zad, built by Ardishir's mother over a dry stream bed. This runs water only when a heavy rain falls; then its water rises like a sea, at which time it extends over a breadth of a thousand cubits and reaches a depth of ninety cubits. The structure, furthermore, narrows as it rises. The gap between the foot of the banks and the bridge, which rises about forty cubits from its base, is filled with a poured out mass of iron slag mixed with lead. The abutments on which the bridge rests, which project from the highest peaks above the bridge, are composed of this substance. Thus, the entire interval between the stream banks and the bridge is filled by a mixture of clinkers and lead. The bridge has but a single arch, a marvel of construction on account of its internal reinforcement.

While clearly exaggerating its dimensions, this description at least recognised the achievement in the construction of this bridge. The extravagant use of lead suggests that the builders had no access to, or perhaps knowledge of, any form of concrete.

Another impressive single arch spanning eighty paces was built in the late seventh century over the River Tab, close to the site of an earlier Sassanid bridge, but the multi-span Sharistan bridge near Isfahan is perhaps the most pleasing early Islamic bridge from the point of view of its structural design. Built in the eleventh century, it blends Roman and Persian elements with its triangular cutwaters upstream and its pointed arches, and a much higher span to pier width ratio than previous Muslim built bridges. It also features unusually large openings over the piers. It could be criticised, as indeed could most Persian bridges and many medieval bridges in Europe, on the basis that the pointed arch strictly demands a concentrated load at its crown to impart stability. This can easily be satisfied in a building, but cannot be satisfied by the transient load on a bridge; but the massiveness of the Persian and medieval European pointed arch bridges, and the relatively small arch spans, ensured their stability in the absence of catastrophic earthquakes, excessive foundation movement or deliberate destruction by military action. The pointed arch did, however, present the advantage that, unlike the Roman semi-circular arch, it allowed the span-to-rise ratio to be varied.

Some of the later Persian bridges became multi-purpose structures, a characteristic they shared with a number of medieval and Renaissance bridges in Europe. The various types of structures that graced these bridges included toll and custom houses, watch towers, pavilions rooms to house guards at border crossings and other rooms to provide rest and shelter for tired travellers. Some structures incorporated both a bridge and a dam.

Two of the most famous bridges of this type are the great bridge of Ali Verdi Khan and the Pul-I-Khaju, both crossing the Zendeh Rud at Isfahan and both seventeenth-century structures. Lord Curzon gives an excellent account of the Pul-I-Khaju, at the same time making some comparisons with the Ali Verdi Khan:

The Pul-I-Khaju is shorter than the Ali Verdi Khan, being only 154 yards [141 metres] in length, owing to a contraction in the bed of the river, which here flows over a ledge of rock. The structure consists, in fact, of a bridge superimposed upon a dam. The latter is built of solid blocks of stone and is pierced by narrow channels, the flow in which can be regulated by sluices. This great platform is broken on its outer edge, the stones being arranged in the form of steps descending to the river level. Upon the platform or dam repose the 24 main arches of the bridge, which is of brick, and chief external features of which are four projecting two-storied hexagonal pavilions, one at each corner, and two larger pavilions of similar shape in the centre, a third storey being erected upon the roof of the more westerly of the two. As in the case of the Julfa Bridge [Ali Verdi Khan] the basement is pierced by a vaulted passage running the entire length of the bridge through the piers on the top of the dam, and crossing the successive channels by stepping stones six feet [1.8 metres] deep. The main roadway of the bridge, twenty-four feet [7.3 metres] broad, is also flanked by a covered gallery on each side, leading to hexagonal pavilions, and opening by a succession of arches on to the outer air.

Finally, there is a terrace walk at the top, originally protected by a double parapet and screens. The pavilions were once adorned with rich paintings and gilding, and with panels containing inscriptions. The decoration is now more jejune and vulgar, and the spandrels of the arches are mostly filled in with modern tiles. In olden days this bridge was a favourite resort in the evening, where the young gallants of Isfahan marched up and down or sat and smoked in the embayed archways overlooking the stream. Now it is well-nigh deserted except in the springtime, when the snows melt in the mountains, and in a few hours the Zendeh Rud is converted from a petty stream into a foaming torrent. Then the good folk of Isfahan crowd the galleries and arches of the bridge, and shout with delight as the water rushes through the narrow sluices, then mounts to the

The seventeenth-century Pul-I-Khaju bridge in Isfahan serves as both a bridge and a dam.

level of the causeway and spills in a noisy cascade down each successive stairway or weir, and finally pours through the main arches, still splitting into a series of cataracts as it leaps the broken edges of the dam.

The degree to which this bridge became a focal point of human activity bears comparison with Old London Bridge and the Ponte Vecchio in Florence. These last two bridges, however, did not serve both purposes as dam and bridge, although Old London Bridge to some extent did so inadvertently, as the narrow arched openings and wide piers raised the water-level enough to cause a flow of water through the arches sufficient to drive water-wheels.

Most masonry bridges built in medieval Europe fell well below the high standards set by the Romans a thousand years earlier and more, both in design and construction. The Romans set a standard difficult to follow, but the medieval engineers and craftsmen certainly had the ability to match them, which they demonstrated in the construction of the great Gothic cathedrals. The marvellous intuitive feel for force distribution and the resulting elegance of structure seen in cathedrals was absent in most of the bridges. Pointed arches provided an ideal structural form for a cathedral, but not for a bridge where the crown of a pointed arch could, in theory at least, develop tensile stresses under bridge loading conditions that masonry could only adequately resist by having thick, massive construction. The stuttering cadences of uneven arch spans, a variety of arch forms, and the cumbersome piers of medieval bridges contrasted markedly with the glorious uplifting rhythms of the arched naves of Gothic cathedrals.

This is not to say that the bridges did not have their own charm; some certainly had this and a degree of structural merit as well. Some of the best and most representative surviving examples of medieval masonry bridges are in Britain, which is appropriate because far and away the most famous and most fascinating bridge built during this period was Old London Bridge. It embodied in generous measure all the typical characteristics, good and bad, of a medieval bridge.

As well as providing a convenient means of crossing a river, many medieval bridges played an important role in the social life of the local community. This resulted in the bridges becoming encumbered with added structures such as dwellings, shops, chapels, toll-houses and fortified towers or gatehouses. As the Church took responsibility for building many of the bridges, it is not surprising that many of them featured a chapel. The bridge priests held regular services in some of the chapels, but more usually they confined themselves to blessing the traveller, collecting monies for the upkeep of the bridge and praying for the souls of the bridge's benefactors. On a medieval bridge in Leeds, a ring of chapel bells summonsed the local clothiers to display their goods on the bridge for the local weekly market.

Of the many originally constructed, only three medieval bridges survive in Britain with original chapels. Two are in Yorkshire, at Rotherham and Wakefield, and one is at St Ives in Huntingdonshire. The St Ives bridge, which like many medieval masonry bridges replaced an earlier timber structure, is still in use despite its narrow 4 metre wide roadway. Its six arches span a total of 61 metres, with four of the original early fifteenth-century pointed arches surviving and two having been replaced by rounded arches in

The St Ives bridge in Huntingdonshire, England, with chapel consecrated in 1426. Four arches, pointed and ribbed, are original fifteenth-century and its two rounded arches date from 1716.

the eighteenth century. The chapel occupies the central pier, its apsidal end projecting well out into the river. A two-storey brick dwelling constructed on top of the chapel in 1736 was removed in 1929 when it became apparent that the chapel walls were giving way under the additional weight. This bridge featured one important medieval innovation in bridge design – the ribbed arch – which the bridge builders almost certainly copied from cathedral construction. The use of ribs supporting a stone shell made possible a lighter form of construction than was possible with a vault having a smooth intrados.

Perhaps the best known medieval bridge in Britain supporting a religious structure is the Chantry bridge at Bradford on Avon in Wiltshire. The nine arches of the bridge (one now buried) present a mixture of pointed and rounded arches, reflecting a history of reconstruction and widening. The pointed arches, strengthened with chamfered ribs, date from the fourteenth century, but were widened downstream in the seventeenth century with rounded arches. A small oratory mounted over a pier near one end of the bridge gave the traveller the opportunity to say a passing prayer before continuing on his way but after the Reformation, at the time of the bridge widening, it was rebuilt with the attractive domed roof added and its use was changed to become a small lock-up. John Wesley spent a night interned there in 1757. A vane with the fish emblem of St Nicholas, the patron saint of travellers, surmounts the roof, giving rise to the saying that a prisoner held there was under the fish and over the water.

Although the many chapel bridges built in Britain had their charm, and indeed some structural merit, they fade into insignificance compared to St Benezet's great structure over the Rhone at Avignon, constructed between 1178 and 1186. According to H. Gautier in his *Traite des Ponts* written in 1714, a group of pious and public spirited men, towards the end of the twelfth century, formed the Frère Pontifes, or Brotherhood of Bridgebuilders, 'to give aid to voyagers, to build bridges, or to establish boats for their use and to receive them in the hospices on the banks of the river'. Although direct evidence of such a pontifical brotherhood is hard to come by, Viollet-le-Duc, the French expert on medieval architecture, romanticised it, relating that a young shepherd named Petit-Benoît came to Avignon under heavenly guidance to build a Great Bridge. After its completion and his subsequent canonisation under the name St Benezet, he became the patron saint of the Frère Pontifes. Having contributed the required money to build a bridge over this difficult site, with the river running wide and deep, the wealthy citizens of Avignon and surrounding provinces would certainly not have entrusted the work to an inexperienced shepherd. They would not have been wealthy if they had been this trusting of divine inspiration. It can be safely assumed that St Benezet, whatever the circumstances, came to Avignon as an experienced engineer and was known to be such. In fact, his previous work seems to have included a bridge over the Durance completed in 1164.

The alignment of the bridge had a sharp elbow and, although the bridge crossed an island in the river, the thirty degrees elbow occurred not on the island, but in the deepest part of the channel. As the elbow pointed downstream, this gives strong credence to Viollet-le-Duc's opinion that the purpose was to give added resistance to the strong flood flows that occur in this river. Twenty-one spans of some 30 metres each made up the total length of 900 metres of the original structure, but only four of these spans survive today. But what magnificent spans! Each arch is made up of four independent stone rings with a distinctive shape, effectively an ellipse with its long axis vertical, apparently arrived at by superimposing the segments of three circles. The rise to half-span ratio of 0.83 is a little

St Benezet's chapel bridge over the Rhone at Avignon, constructed between 1178 and 1186, had twenty-one elliptical spans totalling 900 metres in length. A dog-leg in its alignment pointing downstream may have been to give it added resistance to strong flood flows.

less than the ratio of 1.0 for the Roman semi-circular shape and a better form functionally. Unusually for a medieval bridge, the voussoirs were placed dry without mortar, in the Roman fashion, perhaps influenced by the Pont du Gard, only 30 km away.

The arches sprang from piers 8 metres thick, carried down over 6 metres to bedrock by constructing timber pile coffer dams similar to those used by the Romans. Unlike many of the Roman structures, tapered cutwaters extended downstream as well as upstream, giving a much smoother flow of waters around the piers.

Fortunately the chapel still survives, having been placed near the end of the bridge not swept away by water and ice. In building the chapel, St Benezet took up most of the already narrow roadway of 5 metres, reducing it to 2 metres; the reason for this may have been partly military and partly because the chapel also served as a toll-gate. St Benezet died in 1184, three years before completing his magnificent creation, and his body appropriately was interred in the chapel.

When the papacy moved to Avignon in 1309 to avoid the turmoil in Rome, caused in part by the conflict between the Guelphs (pro-papacy) and the Ghibellines (pro-Emperor), the bridge had already stood for 120 years. The accessibility which the bridge gave to Avignon constituted one of its attractions for the Papacy, as an administrative centre, plus of course the protective arm of the French king. Ironically, a Pope, Boniface IX, committed the first destructive act on the bridge. He ordered some of the arches to be demolished to thwart attacks by troops occupying the defensive tower on the other side of the river. Clement VI had it repaired, but in 1395 it was cut again by a Spanish force and not repaired again until 1418.

Like Old London Bridge, St Benezet's creation has its own popular ditty:

Sous le Pont d'Avignon
L'on y danse, l'on y danse,
Sous le Pont d'Avignon
L'on y danse tout en rond,
Les beaux messieurs font comm'ça
Et puis encore comm'ça.

France also boasts the greatest medieval war bridge still standing. Built between 1308 and 1355 to join the city of Cahors to the limestone hills on the other side of the River Lot, the Pont Valentré presented a formidable obstacle to anyone wishing to enter the city uninvited. Remarkably well preserved, the sharp edges of its cutwaters, square towers and arch rings give it, as Whitney puts it, the aspect of a well disciplined warrior. An immaculately attired French army officer, perhaps. The six lofty pointed arches, each of 16.5 metres span, are striking to the eye, as are the high defensive towers at each end of the bridge and the look-out tower over the middle pier, each rising to 35 metres above low water. Upstream, the high cutwaters present an extended sharp edge to the water current, but downstream they deteriorate to short projections only. Crenellated parapets adorn the tops of the cutwaters.

Shaw Sparrow draws attention to the rectangular holes penetrating the piers and cutwaters just below springing level. Fir saplings thrust into these holes supported planks on which the dressed stones rested before being lifted into place; they also served to support the timber centrings for the arches. Workmen accessed the timber walkways through openings in the piers.

When Raymond Pancelli, Bishop of Cahors, ordered the bridge to be built, he made sure that it was heavily fortified with every known means of defence appropriate against the siege methods of the time. Drawbridges and heavy doors impeded access to the towers at ground level and slits 50 mm wide pierced the walls at first and second floor levels, from which arrows could be fired. Machicolations projecting from the third floor allowed stones or other missiles to be rained down on the enemy, whose advance at this point would have been hampered by a narrowing of the roadway from 5 metres to 3 metres where it passes under the towers. Finally, at the tops of the towers, covered battlements protected the archers as they practised their long-range shooting at the foe.

In Britain, only two war bridges survive: the Warkworth bridge across the River Coquot in Northumberland and the Monnow Bridge on Monmouth. The former has two ribbed segmental arches and, as with many medieval bridges, the outer walls of the triangular cutwaters are carried up to form an integral part of the parapet and provide a refuge where pedestrians could gain protection from mounted riders and wheeled traffic crossing the bridge. It has a small defensive tower at one end. Monnow has a much more imposing tower, sited over a pier, but its semi-circular ribbed arch spans are a little less imaginative in shape. Unfortunately, little is known of the origins of this fine bridge, the archives of the Duke of Beaufort, to whom it belonged, having been destroyed at Raglan Castle in the seventeenth century. When completed in 1272, the roadway was only as wide as the archway in the tower, which hampered the movements of besieging attackers and brought them directly under the row of projections at the top of the tower, from which molten lead could be poured down upon them. The bridge has been widened by springing arch rings from the cutwaters, leaving the original ribbed arches still clearly visible, and increasing the width between parapets from less than 5 metres to 7.5 metres. Openings cut through each side of the tower allow for pedestrian traffic. The bridge has seen military action, having been occupied by both Royalists and Roundheads during the Civil War. In 1839 soldiers occupied the tower in anticipation of an attack by Chartists on Monmouth County Gaol, which never came. They were defeated at Newport.

Other notable medieval bridges in France include the twenty-three arch Pont Saint Esprit over the Rhone, 30 km upstream from St Benezet's bridge; the Pont de Montauban, completed in 1335 across the Tarn; and the picturesque Pont d'Orthez, a thirteenth-century structure with a high octagonal tower over the pier, separating the 15 metre wide arch spanning the river channel from the two smaller approach spans which negotiate the rocky bank on one side of the river. The Frère Pontifes are reported to have built the Pont Saint Esprit, and, like the nearby Avignon bridge, its alignment has a kink upstream. An unusual feature of the Pont de Montauban is the use of red brick for its construction by the engineers, whose names are known – Etiennes de Ferrières and Mathieu de Verdun. Openings above the piers are too high to assist in passing flood flows and may have been provided to relieve weight on the pier foundations and perhaps, also, as a deliberate visual feature of the bridge, or simply a means of saving on building materials.

Monnow war bridge in Monmouth was built in 1272 and later widened. Projections above the gateway allowed molten lead to be poured down on attackers.

Defensive towers originally stood at each end of the bridge. The surviving tower of the Pont d'Orthez once had a twin at the other end of the wide arch. A hole in the parapet walls above the centre of the main span, known as the Priest's Window, proved a useful opening through which in 1560 the Huguenot soldiers of Queen Jeanne could cast into the river priests whom they had taken prisoner. It was at this bridge, too, that forty-five French riflemen, holed up in the tower, halted the English for one day in 1814.

In Italy the unadorned Ponte della Maddelena (also known as the Devil's Bridge) near Lucca, built in 1317, has survived the wild torrential floods of the River Serchio for nearly seven centuries. Its semi-circular main span of 37 metres carries a humped roadway, narrowing from 3.5 metres at the ends to less than 3 metres at the crown.

Many Spanish medieval bridges incorporated the substructures of old Roman bridges, and the superstructures often exhibit a strong Moorish influence. The sixteen-arch bridge over the Gualalqivir River at Córdoba is an eleventh-century Moorish structure resting on the piers of a Roman bridge destroyed in the eighth century. The Puente de San Martin over the Tagus at Toledo, built about 1212, occupies the site of a Roman bridge destroyed by flood, its main pointed arch having a span of 40 metres, the longest of any medieval Spanish bridge. When Archbishop Tenorio ordered the main span to be rebuilt in 1390, it is said that the engineer feared that on removing the centring, the span would collapse. His wife, having prised the information out of him, and realising he was a sorely troubled man, set fire to the scaffolding one night, so destroying the whole work. The husband, now much wiser, went on to complete the work successfully at the second try, an outcome his loyal spouse had presumably anticipated.

Started in the fourteenth century, the famous Karlsbruke (Charles Bridge) over the Moldau in Prague, then capital of Bohemia, took a century and a half to complete. Somehow, the slight S-bend in the alignment adds a piquancy to an evening stroll along its 500 metre length, past the thirty baroque statues placed over the piers in 1707. It crosses the river in sixteen rounded arches ranging in span from 17 metres to 23 metres and, although the piers are provided with substantial cutwaters, separate large timber pierhead structures just upstream of the bridge protect it against drifting ice in the winter time. It is a fitting bridge for a city having claims to be one of the most beautiful in Europe and which, at the time of the construction of the bridge, occupied centre stage in Europe, at the crossroads of east–west and north–south trade routes, making it a prosperous place for craftsmen and merchants. Not surprisingly, its vital location laid it open to many sieges and battles as rival Austrian, Prussian and French armies fought to establish supremacy in Europe. Prague also enjoyed fame as the seat of a university established in the fourteenth century which attracted students from all over the continent.

Old London Bridge occupied centre stage in London for 600 years, its fame deriving not from any outstanding structural merits but because, in addition to providing a rather narrow and congested north–south crossing across the Thames, it became completely integrated into every aspect of city life. As with so many other important medieval stone bridges, it replaced a timber structure at, or close to, the same site; in this case, in fact, many timber bridges preceded the stone structure. In 1999 remains of a massive oak bridge, carbon-dated to the Bronze Age, around 1500 BC, were found some two miles upstream, having a deck 4.5 m wide, sufficient to allow two carts to pass. The finding of a number of artefacts, including a bronze dagger and bronze spearheads, indicated considerable activity at the site and related trade.

The first timber bridge at, or near, the Old London Bridge site is thought to have been either a fixed structure, or perhaps more likely a floating pontoon bridge, thrown across the river by Caesar's military engineers in 55 BC to facilitate his campaign against Cassivellaunus, King of the Catuvellauni. Caesar makes no mention of it in his *Commentaries*, but, as with many of his military bridges, its construction was simply a routine operation, insignificant in relation to his Rhine bridge built the year before. Slightly more positive evidence for a bridge in 43 AD comes from Dio Cassius's *Roman History*, in which he states that some of the German troops under Aulus Plautius, Commander in Chief of the army of Emperor Claudius, swam across a tidal lake which the Britons they were pursuing crossed more readily with fore knowledge of firm ground and easy passages. But others of the Germans '…got over by a bridge a little way upstream, after which they assailed the barbarians from several sides at once…' More compelling evidence of a Roman bridge came to light during dredging of the Thames to deepen its channel after the removal of the Old London Bridge piers in 1832. The gravel and silt removed from the river bed contained many thousands of Roman coins and medallions, probably offerings to the river gods to ensure safe travel. Remains of oaken piles with iron shoes also pointed to the existence of a Roman timber bridge at this site and, indeed, there may have been a succession of such bridges over the four centuries that Britain remained a province of the Roman Empire.

There are no known historical records of a bridge for several centuries after the departure of the Romans. Stowe makes reference in his annals to a ferry and Smiles tells of a 'traditional' story associated with a ferry operated by one John Overy, who grew rich on its proceeds. He was also very mean. During one of his fits of usury he decided to feign death for twenty-four hours in the expectation that his servants would fast during this period, thus saving him money. Unfortunately for him, they took the opportunity instead to break into his larder and when he rose in his winding sheets to protest, one of the servants brained him with an oar, fearing him to be the Devil. A gallant who sought John Overy's beautiful daughter Mary in marriage, hearing the news, rode posthaste into town from the country, perhaps in order to inherit the prosperous ferry business, but fell off his horse and 'brake his neck'. Mary assuaged her bottomless grief by founding the church of St Mary in Southwark and made over all her possessions to it.

References to a London bridge again appear in the records dating from the tenth century, including the drowning of a witch at London Bridge and payments of a toll for boats coming up to the bridge. A bridge is also mentioned in the accounts of the attacks on the city in the early eleventh century by Svein Fork-Beard, the Danish King. Shortly after, the Danes themselves came under attack by the Norwegian Olaf who, after erecting protective canopies over his ships to protect them from missiles, rowed them up to the bridge, fastened ropes around the supporting piles, and rowed back down on the retreating tide, so demolishing the whole structure. It was chronicled by the Norse poet Ottar Swarte in the *Olaf Sagas*:

> London Bridge is broken down,
> Gold is won, and bright renown.
> Shields resounding,
> War-horns sounding,
> Hildur shouting in the din!
> Arrows singing,
> Mailcoats ringing –
> Odin makes our Olaf win!

This song later gave rise to the famous nursery rhyme which appeared in 1810 (although it is much earlier in origin) in *Gammer Gurton's Garland* or the *Nursery Parnassus*, a 'choice collection of pretty Songs, and Verses, for the amusement of all little good children who can neither read nor run':

> London Bridge is broken down,
> Dance o'er my Lady Lee;
> London Bridge is broken down,
> With a gay lady.

There then follow many verses recounting possible methods for rebuilding it, and their drawbacks: silver and gold would be stolen away, iron and steel would bend and bow, wood and clay would wash away. But there was an answer!

Build it up with stone so strong,
Dance o'er my Lady Lee;
Huzza! Twill last for ages long,
With a gay lady.

In 1091 the then existing timber bridge succumbed to a flood, but William Rufus levied a heavy tax for its replacement, which succumbed in turn, but to fire, fifty years later. The constant patching up and rebuilding required to keep a timber bridge operational at the site, added to the perils of flood and fire, led eventually to the decision in 1176 to replace it with a stone bridge. Peter, the Chaplain of St Mary Colechurch, carried out the final work on this last timber bridge, built of elm, before taking on the commission to build the new stone bridge close by. Obviously a very competent engineer, his work on the timber structure no doubt made him realise the inadequacy of the timber structure, requiring constant attention, serving a city of rapidly growing importance. His arguments clearly impressed both his Church leaders and Henry II, the former donating substantially to the new structure and the King imposing a tax on wool to help pay for it, giving rise to the legend that the foundations of the bridge rested on woolpacks.

Completed in 1209, the bridge took thirty-three years to build, construction of each arch taking an average of nineteen months. Peter of Colechurch did not see its completion, having died in 1205. King John, presumably concerned by Peter's increasing infirmity, had previously recommended to the mayor and citizens of London that they should employ the French engineer Isenbert to complete the bridge, his previous work having included important bridges at Saintes and La Rochelle. In the event this recommendation was not taken up, and when Peter died the responsibility for completing the structure fell to three worthy merchants of London. Appropriately, Peter's remains were interred in the undercroft of the bridge chapel, which he himself had dedicated to Thomas Beckett, who had been heinously murdered in his cathedral at Canterbury in 1170 at the instigation, inadvertently or otherwise, of Henry II. This tragic event apart, Henry proved himself an outstanding king, ruling over large areas of France as well as Britain, where he strengthened the legal system, imposed taxes and forced law and order on the kingdom after the disastrous reign of Stephen, and his support for the building of the new bridge can be seen to fit well into this context.

Arch spans ranged from 4.5 metres to 10.2 metres and pier widths from 5.5 metres to 8 metres, making up a total bridge length of 276 metres. The range of spans may be explained, in part at least, by the use of pointed arches, with their flexible rise to span ratio, leading to a rather casual attitude in siting the piers. The seventh span from the southern end of the bridge was a timber drawbridge opening, nearly 9 metres in width. The same haphazard array of arches can still be seen today in the twenty-four-span Bideford bridge across the Torridge River in Devonshire. Completed in 1315, a century after Old London Bridge, with no two spans alike, it gives a good idea of the appearance of London Bridge without its incumbent structures, despite having been widened and substantially rebuilt in the nineteenth and twentieth centuries.

The uneven spacing of piers for Old London Bridge and other medieval bridges may have owed something to the extremely difficult problem of driving piles into the river

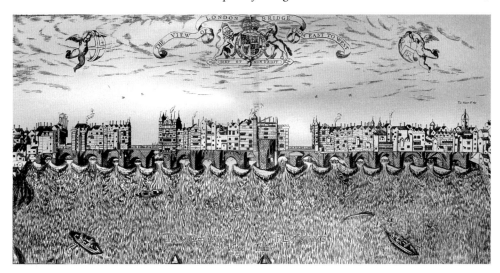

London Bridge from an engraving by John Norden, about AD 1600. Completed in 1209, it experienced many repairs and partial rebuilds in over 600 years of life and the incumbent structures underwent a number of transformations.

bed to support them. This would probably have been done from a barge or other floating vessel, held in position perhaps partly by drag anchors and partly by cables fixed to the vessel at one end and lashed at the other end to piles, either driven already for the bridge or driven specially for the purpose of holding the vessel in place. In a tidal river, such as the London reach of the Thames, the vertical movements of the floating vessel may have been accommodated by constant adjustments of cable lengths, or by simply allowing sufficient slack to take the movements. In either case, it would have been difficult in the extreme to position the vessel and its pile driving rig and hold its position against the river flow and tidal currents that varied both in strength and direction. It is possible that fixed platforms could have been used, but these would have been major structures themselves founded on piles, and thus would have presented their own positioning problems.

Demolition of the old bridge in 1832 enabled a study to be made of the original foundations. These consisted of closely spaced elm piles driven deep into the bed of the river, with large stones packed into the interior spaces between the pile tops, which terminated at about low water level. Strong oak sleepers bedded transversely across the tops of the piles supported the bases of the piers, the lowermost layers of the squared stone blocks being bedded in pitch. The pier widths totalled 123 metres, taking up slightly less than half the river width, thus greatly increasing the flow velocity of water through the bridge and consequently increasing the possibility of scour undermining the bridge foundations. In order to protect the foundations from such scour, protective starlings were built around each pier base, composed of rows of closely spaced piles with stone rubble infilling enclosed within a boat-shaped ring of contiguous elm piles. A grid of timber planks bolted to the tops of the piles brought the elevation of the starlings up to about half-tide level. This further restricted the width of river flow up to half-tide to about 70 metres, or little more than one-quarter of the original river width. The restriction to the flow caused an average drop in water level along the lengths of the starlings of 0.6 metres at neap tides, 1.3 metres at spring tides and as much as 1.7 metres for conditions of a land flood and spring ebb-tide. Inevitably, this led to the starlings suffering constant damage and needing repairs, each time cribbing a little more of the river width. Nevertheless, the bridge survived over 600 years before demolition, attesting to the effectiveness of the starlings in fulfilling their purpose.

Throughout its existence Old London Bridge underwent many changes of appearance as its various incumbent structures came and went, standing in all their glory until their decay or destruction by fire or other agencies, to be followed by demolition and replacement. The bridge supported these structures until 1760, when it was deemed no longer fit to do so and they were demolished, the bridge itself meeting the same end in 1832. With a roadway width of only 6 metres, the houses and other incumbent structures had to be cantilevered out over the water, supported by inclined timber props, the like of which can still be seen today in the Ponte Vecchio in Florence. The extent to which the bridge permeated the very hearts and souls of London people is reflected, among other things, by the fact that over the years most of the arch openings assumed their own distinctive names. Those acquainted with the bridge knew well Rock Lock, Pedler's Lock, Gut Lock, King's Lock and other similar appellatives.

Little is known of the structures placed on the bridge at the time of its construction or shortly after as no drawings of this period have been found. Certainly the dominating features included the chapel and the two gates, the outer of which, the Great Stone Gate, stood over the second pier from the southern, Southwark, end. It collapsed in 1347, bringing down two arches with it. Its replacement survived until it was badly damaged by fire in 1725, when it was rebuilt only to be demolished in 1760. The presence of this fortified tower proclaimed to all potential enemies of London, approaching from the south, the city's resolve and capability of protecting itself, but the bridge never experienced military action. The timber drawbridge spanning the second opening from the southern end, together with its gate tower occupying the northern pier of the opening, gave a second line of defence, which again proved unnecessary; but it served for four hundred years its primary purpose of allowing tall masted ships to sail past the bridge to Queenhithe, the bridge wardens imposing a toll of 6d to raise the span. The original drawbridge gate housing the chains and mechanism to raise the span may have been of timber. It was replaced in 1428 by a stone structure which served for 150 years, when it too gave way to the most remarkable structure to grace the bridge – Nonesuch House.

Unlike the two gate towers, the chapel did not straddle the bridge, but stood on one side of it on an extended pier, projecting nearly 20 metres beyond the side of the bridge. A two-storey structure, it could be entered directly from the roadway at the upper level, and from the starling at the lower level; this latter facility allowed fishermen access after tying up their boats to the starling and gaining access by a set of steps to the lower storey of the chapel. Towards the end of the fourteenth century Henry Yevelle, one of the wardens of the bridge, recommended the demolition of the old chapel. It may well have been in poor condition, but equally Yevelle may have fancied the idea of building a chapel on the bridge to his own design. Funds appear to have been generously available from benefactions, endowments and donations, the dedication to St Thomas Becket helping greatly in this regard. Regrettably, it was also this dedication that incensed Henry VIII who, in the process of dissolving the monasteries, ordered all representations of the saint throughout the realm to be expunged. There seems to have been some unwillingness to comply with this order to desecrate Yevelle's chapel, which occupied a special place in the hearts of London people, and it was not until 1553, six years after Henry's death and towards the end of the brief reign of the young Edward VI, also a fanatical supporter of the new Protestantism, that the chapel was defaced and turned into a habitation and grocer's shop.

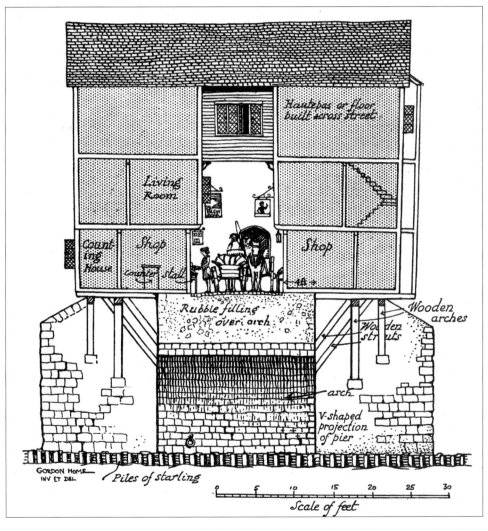

A sectional drawing by Gordon Home of Old London Bridge around 1500, showing an arch and pier to scale and the probable arrangement of houses and shops. The roadway averaged less than 4 metres in width.

Houses and shops appeared on the bridge shortly after its construction, but within four years a large fire broke out in Southwark and spread rapidly, engulfing a number of houses at the southern end of the bridge. Very quickly, wind-borne burning debris carried to the northern end of the bridge and set fire to houses there, trapping people on the bridge. Panic ensued. Many people lost their lives trying to escape through the flames, leaping over the parapets, or jumping into boats which, already overcrowded and made unstable by the rush of water through the restricted arch openings, capsized, spilling the occupants into the turbulent river.

Fire remained a constant threat to the wooden structures on the bridge, but despite minor outbreaks no further major conflagration occurred until 1632, when a maidservant in the employ of a needlemaker, Mr John Briggs, at the northern end of the structure placed a tub full of hot coal ashes under the stairs before going to bed. The ensuing inferno destroyed forty-three houses occupying the northern third of the bridge. By 1651 a much admired integral block, rather than individual dwellings, had been completed, filling about one-quarter of the gap left by the fire. Mercifully, the remainder of the gap still remained to protect the bridge when, in 1666, the Great Fire of London broke out, destroying the new block but leaving the rest of the bridge unscathed. The fire also destroyed waterwheels that had been placed in the arch openings under the new block. By 1683 most of the northern end of the bridge again groaned under the weight of three-storeyed structures along its length and with the replacement, too, of most of the houses at the southern end, the bridge lost much of its medieval character, other than its own basic structural form of uneven pointed arches, somewhat out of harmony with the regularity and uniformity of the buildings it now supported.

28 August 1577 saw the laying of the foundation stone for Nonesuch House, the most famous building to be erected on the bridge. It occupied the site of the famous Drawbridge Gate. Its unlikely name derived perhaps in part from its exotic appearance and in part from having been built entirely from timber frame units prefabricated in Holland and joined together on site with wooden pegs, reputedly with not a single nail being used in its construction. On the outside its brilliantly painted woodwork was bedecked with ornaments, its four corners turreted and topped with onion-shaped cupolas and golden vanes, its carved and gilded gables supported by scrolls. For a time it housed nobles of Queen Elizabeth's court, but little is known of its subsequent occupants up to the time of its demolition with the remainder of the houses in the eighteenth century.

Notwithstanding constant noise from human and animals, carts lumbering across the structure, the roar of the river through the arched openings and the tortured groanings and creakings of the waterwheels, the bridge was regarded as a desirable and healthy place on which to live. Unlike most medieval streets, rubbish went through the trapdoors into the river and not onto the street. At the time of the Black Death in 1348–9, only two inhabitants of the bridge died. Most of their daily requirements could be obtained from shops lining the two sides of the narrow roadway, the vendors displaying their goods on hinged boards let down during the day to project into the roadway, further restricting its width, and raised at night to act as a shutter. A decree passed in 1580 forbade the use of counters projecting more than 4 inches beyond the shop front. Medieval records refer to a number of shopkeepers operating in this way, mostly selling goods made on the premises. These included a glover, goldsmith, cutler, haberdasher, armourer, bowyer and fletcher (bows and arrows) and a tailor. Butchers, grocers and fishmongers either traded on the bridge or in the nearby lanes at either end of the bridge. In keeping with the inhabitants, the shops later moved upmarket, a survey taken in 1633 revealing that thirty-two of the thirty-eight shops on the bridge sold goods in some way connected with wearing apparel, including milliners, hosiers, glovers, needlemakers and silkmen. Booksellers, too, now plied their wares on the bridge.

After Nonesuch House, perhaps the most famous building associated with the bridge was the Bear Tavern, hard against the Southwark shore; its fame owes something to Samuel Pepys, who frequented it, as no doubt did many of those who lived or worked on the bridge or close by. After a few convivial drinks with friends, clientele could repair to one of the many profitable whore-houses in Southwark owned by the Bishop of Winchester. It is safe to assume that the Bear also enjoyed the patronage of the wardens and clerks who conducted the business affairs of the bridge from the nearby Bridge House Estates, sited on the Southwark shore. They collected rents and tolls, which, together with donations and bequests for the upkeep of the bridge, accumulated to such an extent that Bridge House Estates counted among the wealthiest landowners in London. The Estates made substantial financial contributions towards the cost of constructing other London bridges, including Old Blackfriars (1769), the New London Bridge (1831), which replaced Old London Bridge, and Tower Bridge (1895). It also provided funds for the purchase in 1868 of Southwark Bridge from the private company which had constructed it in 1819 as a toll bridge. The toll-houses were removed after the purchase.

Storehouses and yards attached to Bridge House contained vast quantities of paraphernalia required for the upkeep and repair of the bridge. Timber enough for fourteen complete shops, and 120 elm pines for repairs to starlings, appeared in the 1350 inventory, together with 1,734 blocks of Portland stone, as well as Kentish ragstone and chalk, and huge quantities of rope, lead and nails. Implements and equipment included pile drivers, rakes for catching floating debris, cauldrons for melting pitch, two boats and a barge. Occasionally the plant seems to have been let out for hire by the wardens.

Bridge House and adjacent buildings also provided space for the storage of large quantities of grain for the admirable purpose of providing wheat to London bakers at time of dearth at a fixed and moderate price. Waterwheels at the southern end of the bridge drove mills for grinding the corn. Various troubles arose over these granaries, including complaints by bakers compelled to buy the corn that despite regular turning and screening to keep it fresh, it had become mouldy. As they could be convicted and even pilloried for selling bad bread, this constituted a serious matter for them.

On a number of occasions in the life of the bridge, severe winter weather conditions led to freezing up of the Thames, notably in 1281, 1564–5, 1608, 1683–4, 1715–6, 1739–40 and 1814. Lumps of ice floating down the river blocked the arch openings, banking up the water behind, which then froze. With such an opportunity beckoning, Londoners took to the ice, some to skate and some 'plaied a the foot-ball as boldlie there as if it had been on drie land', according to one temporary account of the 1564–5 frost. Later, severe frosts saw great fairs held on the ice, the frozen surface of the river supporting various stalls and carriages. Printing presses sprang up. The painter William Hogarth, who lived on the east side of the bridge for a time, visited the fair of 1740 and it is recorded that he stopped at one of the printing booths where, for a few pence, he had not his own name printed (a popular souvenir of the occasion) but that of his dog Trump. Other diversions included sledge races, horse-and-coach races, puppet plays and even bull baiting. Whole oxen were roasted on the ice and the merriment went on through the nights as well as the days.

'Shooting the Bridge' consisted of passing through the arch openings in a small boat, particularly at times of spate when the roiling cataracts under the bridge reached their peak, presenting an irresistible challenge for those looking for excitement and danger, and it claimed many lives. The popular saying that 'London Bridge was made for wise men to go over and fools to go under' did not deter a number of famous men, including Pepys, and some women from taking it on. A great sensation followed the death in 1689 in this manner of the recently appointed Secretary of War, John Temple. He hired a small boat with its attendant waterman, whom he instructed to shoot the bridge. Unknown to the waterman, Temple had filled his pockets with stones and on reaching the cataract he flung himself overboard and drowned. His depression arose from the fact that he had had his friend Captain Hamilton, who had been incarcerated for high treason, released from the Tower on the agreement that he (Hamilton) would go to Ireland to try and persuade the Earl of Tyrconnel, leader of the Irish rebellion, to submit. Hamilton chose instead to join Tryconnel and led a regiment against the English. On a more happy occasion Queen Elizabeth I passed under the bridge on a ceremonial river journey three days before her investiture, accompanied by the Lord Mayor and City Companies and 'with great and pleasant melody of instruments', sensibly negotiating it at the ebb of the tide. A little over four years earlier she had passed under the bridge on her way from Richmond Castle to Woodstock as a prisoner of her half-sister Mary, on which occasion she would have seen the heads of some of her supporters on the blood-stained poles atop the Drawbridge Gate.

The grisly and barbaric practice of exhibiting traitor's heads on the bridge continued for nearly 400 years. Originally displayed on poles on the Drawbridge Gate, they were simply transferred on the demolition of that structure in 1577 to the Great Stone Gate. The first known instance of a head being placed on the bridge was that of William Wallace, executed in 1305, and the last head was that of William Stayley, executed in 1678 for his involvement in the Popish plot. In between, many individuals shared this fate, from common rebels like Wat Tyler to men of high birth or distinction such as Sir Thomas More, executed in 1535 for not accepting Henry VIII as head of the Church. His daughter removed his head after a few months, but most heads remained until they rotted. In some cases, the clumsy attempts at severing the heads of hanged individuals resulted in part of the upper trunk as well as the head appearing on the pole. Sometimes, too, the poles received a batch of heads all at once, such as eight of the Gunpowder Plot conspirators in 1605. A leak by one of the conspirators of their plan to blow up the Houses of Parliament at the opening by James I, in an attempt to promote Roman Catholicism, led to the discovery of Guy Fawkes in the cellars on the point of firing the 13 tons of gunpowder.

Although Old London Bridge never experienced any military battle, it did witness many little vignettes of history, some of lasting significance, others more transient. An early incident concerned Wat Tyler at the head of the Peasants' Revolt in 1381. When the 14-year-old Richard II imposed a swingeing poll-tax, requiring even the humblest peasant to pay 3 groats, Kentishman Wat Tyler led a group of insurgents in a march up to London to voice their objections. Temporarily checked by the Mayor raising the drawbridge, they threatened to burn down the houses on the bridge, whereupon the citizens, not unsympathetic to the peasants' cause, lowered it. They then rampaged through the city, venting their spleen on distinguished citizens and foreigners, killing a considerable number. Simon of Sudbury, Archbishop of Canterbury and Chancellor, and the Treasurer Robert Hailes, were taken and executed, and their heads, together with eight others, were placed on poles on the Drawbridge Gate. The youthful Richard arranged a meeting with Tyler at Smithfield, which went well at first, with Richard granting most of the peasants' demands. But suddenly, his blood up, Tyler launched an untimely tirade at the King and his counsellors; weapons were drawn and Tyler struck down, after which the crowd dispersed. As they streamed back across the bridge, they must have been all too aware of Wat Tyler's head atop the Drawbridge Gate, together with that of Jack Straw, the Essex rebel leader.

In 1480 another Kentishman, Jack Cade, led a rebellion not of peasants but of middle-ranking citizens, against the inefficient and oppressive officialdom. The citizens opened the bridge gates for him to cross, but soon regretted it as he carried out indiscriminate attacks on their property and persons, including setting fire to houses at the southern end of the bridge. A fierce battle on the bridge ensued that Cade lost, and as a result he retreated back to Kent. He died of wounds after being captured and his head found its way to the usual place, as did those of many of his followers or supposed followers.

Ostensibly to free King Henry VI from the tower, but more likely to enrich himself by plundering the city, Thomas Fauconberg, known as the 'Bastard Fauconberg', in 1471 laid siege to the bridge at the head of a group of desperadoes. The worthy citizens, led by Ralph Jocelyn, alderman and draper, overcame the insurgents and more heads, including that of Fauconberg, found their way to the top of the Drawbridge Gate.

Perhaps the most poignant moment occurred on 4 December 1586, when the sentence of death passed on Mary Queen of Scots was read at London Bridge in the presence of the Lord Mayor and various nobility and leading citizens of London, bedecked in velvet and gold chains and all mounted on horseback.

In his *Ten Books of Architecture*, published in 1485, Alberti recommended the use of the semi-circle as the strongest of all arches, but conceded that if 'the semi-circle should rise so high as to be inconvenient, we may make use of the Scheme Arch, only taking care to make the last piers on the shore the stronger and thicker'. By a Scheme Arch he probably meant a segmental form, with which he would have been quite familiar, as the Ponte Vecchio in Florence, with its flat segmental arches, had already seen a hundred years of service when Alberti completed his manuscript in 1452. He clearly recognised the much greater horizontal forces exerted by the flatter form of arch, and that these could be carried through the bridge to the abutments. It was another three centuries before the outstanding French bridge engineer Jean Rudolphe Perronet exploited this principle to the full. Alberti recognised that both the piers and arches should be built of the hardest and largest stones and that the thickness of the arch ring should be at least one-tenth of the arch span, which itself should not be more than six times, nor less than four times, the thickness of the piers. The stones should be fastened together with pins and cramps of brass.

He places particular stress on the need to provide foundations and piers able to resist the scouring action of the water and advises constructing the foundations within an enclosure (a coffer dam) to keep out the water, either digging down until a solid strata is reached or driving piles with burnt ends. He recommends constructing the coffer dam by driving a double row of closely spaced stakes projecting above water level enclosing the working area, and making it water tight by ramming in reeds and mud to fill the gap between them. The piers themselves, he says, should be higher than the fullest tide level, with a thickness one-fourth the height of the bridge, and should have an angular end at the stern (downstream end) as well as at the head (upstream end) as eddies at the stern could be a greater cause of scour than the water impacting on the head. He suggests three-quarters, or possibly two-thirds, of a right-angle as being a suitable angle.

When Taddeo Gaddi (a pupil of Giotto) commenced his construction in 1345 of the Ponte Vecchio (Old Bridge) over the River Arno in Florence, he knew all too well that several previous bridges at the site, both timber and stone, had been swept away by thunderous floods sweeping down the river channel. Only twelve years earlier the most recent bridge at the site, together with two other bridges along the river, the Trinita

Taddeo Gaddi commenced construction of the Ponte Vecchio in Florence in 1345. The two-storey structure of shops and passageway it supports almost masks the thinness and elegance of the arch rings. The flat segmental arches have an unprecedented rise to half-span ratio of 0.31 and the pier widths of one-fifth the span compare to Roman bridges of about one-third. It has survived the rampaging Arno since its construction.

and Carraia, had been destroyed by a flood that claimed 300 lives. Gaddi's picturesque bridge still survives today, having withstood many floods. The other two bridges, having been rebuilt, again succumbed to the great flood of 1557. On 4 November 1966 the Arno produced its worst ever flood after prolonged heavy rain and an untimely discharge of water from a hydro-electric dam upstream. The flood did vast amounts of damage as it broke its banks, allowing water to rampage through the city; but, as always before, Gaddi's bridge held, although some of the buildings on it suffered severe damage.

Taddeo Gaddi's solution to the problem comprised digging his foundations deep and to have pier widths as narrow, and spans as wide, as he dared, thus causing minimum restriction to river flow. The 6 metre pier widths are only one-fifth of the central 29 metre span, which contrasts with the one-third ratio common in Roman and medieval bridges. As semi-circular arches with such large spans would have raised the bridge roadway much too high, Gaddi adopted segmental arches, the central span having a rise of 4.5 metres, giving an unprecedented rise to half-span ratio of 0.31.

Another factor giving the bridge stability is its generous width of 32 metres, allowing shops to be built on both sides of the roadway. Butcher's shops existed on the bridge for nearly 200 years, but Cosimo I ordered them off in favour of the goldsmiths, silversmiths and jewellers. These businesses created additional space for their operations by extending the backs of their premises out over the water, supported on timber props, contributing to the rather disordered and picturesque appearance of the bridge today. Most of the shops are still jewellers, interspersed with leather goods and clothing accessories. The Grand Duke Cosimo commissioned Vasari to build the covered passageway above the shops on the east side of the bridge so that he could ride without getting wet from his official residence at the Palazzo Pitti to the Uffizi, originally built by Vasari to house the administrative offices of the Medici state. It is now a famous art gallery.

The Ponte Santa Trinita, just downstream from the Ponte Vecchio and also destroyed in the 1333 flood, was rebuilt a year after its upstream neighbour, but did not survive the 1557 flood. This gave sculptor turned civil engineer Barttolommeo Ammanati the opportunity to build a bridge with arches of a shape which mathematicians ever since have argued about and attempted unsuccessfully to define in mathematical terms. In defiance of recommended good taste for the time, he adopted pointed arches for the three spans, albeit pointed only shallowly at the crowns. This feature cannot be seen as the arch crowns are hidden behind carved escutcheons or pendants. The tangents to the soffits are vertical at the springings, the subtle curvatures at this point giving the impression of the arch rings merging seamlessly into the piers. The three arch spans of 26 metres, 29 metres and 26 metres are almost identical to those of the Ponte Vecchio, and the rise to half-span of 0.29 also corresponds closely to the adjacent bridge, but the shapes of the arches could hardly be more unalike. The curves have been compared to compound ellipses, reversed parabolas and various other shapes. In his youth Ammanati had studied drawing and, as an artist, he probably sketched freehand several forms of curves until he had on paper shapes which pleased him and gave him a warm feeling, which he then transferred to the bridge.

Fully cognisant of his own limitations and relative inexperience as a civil engineer, Ammanati hired Alfons and Guilo Parigi, father and son, to act as engineers-in-charge of construction. Many details of their works have been preserved in a notebook they kept, including sketches of the timber centring for the arches and of the foundation works. To enable the foundations to be constructed in the dry, they isolated each pier position by a cofferdam 46 Braccia (27 metres) by 14 braccia (8 metres) in plan, consisting of double rows of timber piles, with 4 braccia spacing between them, filled with a lime and gravel mortar. After pumping out the water, the soft material in the bed of the river was excavated to a depth of 7 braccia, and piles 7 braccia long driven, probably to refusal. Masonry bed courses consisting of large flat stones some 2 metres long and 0.5 metres wide were laid on top of the piles to support the piers. The prodigious amount of attention paid to the foundation works clearly reflected the concern felt by Ammanati and the Parigis at the destructive force of the Arno in full spate, with debris-laden flood waters sweeping along at over 60 km/hour. Even they could not reckon with the retreating German army needlessly destroying the bridge during the Second World War.

That the Ponte Santa Trinita stands today, identical in form to Ammanati's bridge completed in 1570, is a tribute to the affection of the Florentines for this structure. As far as possible, the restoration after the war was carried out with the original stones salvaged from the river bed. Francavilla's seventeenth-century statues, representing the four seasons, which graced the bridge were recovered from the river, but not the head of Primavera. A widespread search conducted not only in Florence, but in Germany and throughout the world in case someone had carried it off, revealed no trace of it until, one day in 1961, a workman happened upon it while walking along the river bank. It made headlines in the Florentine papers: '*E Tornate La Primavera!*' – Spring has returned.

As the Rialto Bridge in Venice only had to span the Grand Canal, the designer Antonio da Ponte did not have the problem of fast flowing floodwaters to contend with, but he did have to provide suitable foundations for the abutments to cope with the notoriously poor ground conditions, as the area was originally a swamp. A thirteenth-century pontoon bridge may have been the first fixed crossing at the site, succeeded by a number of timber structures, each with a life limited by decay or fire. The last of these structures had a central double bascule span which could be lifted to allow through the great ships which took part in Venice's Ascension Day procession. Each year the reigning Doge sailed into the Adriatic in the great state barge, the Bucentaur, and threw into the water from its deck a consecrated ring to symbolise the marriage of Venice to the sea.

Long before the commencement of construction of the present bridge in 1587, it had been realised by responsible citizens in Venice that sooner or later a permanent stone bridge would have to be built. An opportune moment to push hard for it came when a great fire in 1512 destroyed many buildings in the Rialto district and threatened the timber bridge. In the event, it escaped harm and subsequently procrastination won the day. Many famous engineer/architects prepared plans for a stone structure, including Giacondo (who had built the Pont Notre-Dame in Paris), Michelangelo, Vignola, Sansorino and Palladio. Only Palladio's drawings survive. All the designs were ignored, despite the great prosperity Venice enjoyed, her ships sailing the Mediterranean and beyond, trading in wool, metal, leather, furs, lumber and soap – for spices, perfumes, fabrics and slaves. Instead, prosperity begat buildings that impinged ever closer to the bridge site, making increasingly difficult the task of constructing foundations capable of supporting a single arch stone bridge across the Grand Canal.

At last, in 1587 the Venetian Senate decided that a permanent stone bridge should be built, the existing timber structure having no doubt deteriorated to a condition both dangerous and beyond repair. A Commission of three of its members, set up to select a plan and take charge of construction, gave serious consideration to designs submitted by two engineer/architects of good repute in the building profession, but inexperienced in stone bridge design. One of these, Vincenzo Scamozzi, then 35 years of age, with the encouragement of his engineer father had made a study of the buildings in Venice before going to Rome to take up mathematics under a famous teacher there. He returned to Venice in 1583 and immediately received several minor commissions. The other designer was the appropriately named Antonio da Ponte, Curator of Public Works in Venice and already 75 years of age. His fame, modest at the time, stemmed not from any great architectural or engineering achievement on his part, but from his own personal efforts in saving the Ducal Palace after it suffered two damaging fires in 1574 and 1577. He not only took part personally in extinguishing the fires, but fought hard against Palladio, among others, to have the building restored rather than demolished and rebuilt after the 1577 fire. His views prevailed and the world owes him a debt. It also owes a debt to the Commission for choosing Antonio's magnificent single span design ahead of two designs by Scamozzi, one of three spans that he himself preferred and the other of one span. The Commission rejected Scamozzi's plans mainly because his proposal to construct the pier foundations for the three-span bridge would have blocked the canal to traffic for two years, while his abutments for one span, founded on well rammed ground, would have been inadequate to cope with the thrusts from the bridge arch.

Antonio da Ponte's simple segmental arch has a span of 27 metres and a rise of 6.4 metres, sufficient to pass all state barges. Its generous width of 23 metres accommodates two rows of shops with a centre passageway and two sideways. As a minor quibble, the design can perhaps be criticised on the aesthetic grounds that the incumbent structures and balustrade tend to obscure the slender gracefulness of the marble arch, with a thickness of only 1.4 metres at the crown. Wide, shallow steps carry the roadway over the arch.

Da Ponte solved the foundation problem in a unique way, excavating into the canal banks for each abutments in a series of three steps, with the shallowest level at the back to cause minimum disturbance to nearby buildings, and the deepest level, taken down to 5 metres depth, at the water's edge. The excavations were made in the dry inside cofferdams, a task he had to take over himself as the contractors proved incompetent. At each abutment site, 6,000 birch-alder piles, 150 mm diameter and 3.5 metres long, were driven to refusal, then cut flush with the excavation step level and capped with grillages of larch. Above this, da Ponte placed a layer of wedge-shaped stones, backed with bricks, to provide a base enabling him to build up the abutment to ground level with layers of stone angled in an attempt to incline the main joint planes at right-angles to the line of thrust from the arch. This inclination of the masonry joints continues into the visible spandrels of the bridge.

Antonio da Ponte completed construction of the 27 metre span Rialto Bridge in Venice in 1591. It survived a strong earthquake shortly after construction.

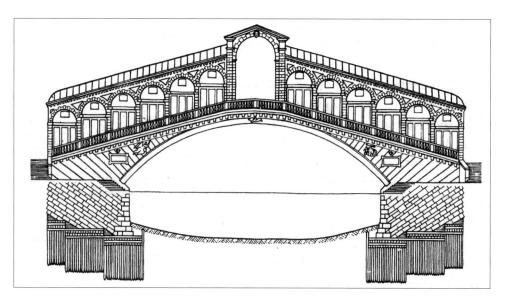

Rialto Bridge, showing a section through the foundations. Antonio da Ponte stepped his piled foundations to avoid damage to nearby buildings and angled the abutment stones to better resist the inclined arch thrusts.

In addition to the difficult construction problems confronting him, da Ponte had to endure during the course of the work rumours and criticisms regarding the adequacy of the structure and its foundations, some of which may have originated with Scamozzi and his friends, who appear not to have taken with good grace his loss of commission to design the structure. The three curators appointed by the Senate to oversee the work could not agree between themselves on the adequacy of the foundations and one of them, Alvise Zorzi, a friend of da Ponte, wrote to the Senate recommending that they appoint a new Commission to report on the work. As the rumours had led to considerable public disquiet, the Senate agreed to this and directed the new Commission to report within six days. In two sessions on 11 and 12 August 1588, before the special Commission, fourteen persons with some expertise offered opinions on the adequacy of the completed foundations. Most declared in favour of the work. The excavation of stepped rather than level foundation bases drew some criticism, as did the placement of abutment masonry at an inclination rather than horizontally; but these and other minor criticisms were convincingly countered by Alvise da Ponte, a relative of Antonio, who had been in charge of constructing the foundations. In the debate that followed, Antonio da Ponte defended his design with clarity and dignity, strongly supported by Zorzi. Despite some convoluted arguments by Scamozzi's supporters, the Commission approved da Ponte's design, including the work which had been completed, and the Senate gave him the necessary authority to complete the work.

Final vindication of Antonio da Ponte's design came in a rather unwelcome way in 1591, shortly after completion of the bridge. On 10 July a strong earthquake shook Venice and several other Italian cities. Shopkeepers started to flee from the bridge when it, like other structures in Venice, started shaking, fearing that it was about to crumble, but returned no doubt with their belief in the structure strengthened when it proved not to have suffered any damage, not even a single crack. It must have brought a great sense of satisfaction to the old man, now 79 years of age.

Paris had its origins as a fishing village on an island in the Seine now known as La Cité. It became important in Roman times after Julius Caesar established there the Sovereign Council of the Gauls, making it the seat of Roman authority. A succession of early timber bridges, during and subsequent to the Roman period, connected the island to both banks of the river: as each one fell into decay or became the victim of flood or fire, a new structure replaced it. The final and best known timber structure, completed in 1416, linked the island to the north shore and gave access to the cathedral of Notre Dame, which in turn lent its name to the bridge.

As the monks of St Magliore claimed exclusive rights to this portion of the river, the city authorities purchased from them title to a width across the river of 12 loises (23 metres), but they cribbed a bit, building the bridge with a width of 29 metres to allow two rows of houses to be accommodated on it. The monks expressed no concern at the additional width, but claimed that rentals from the houses, occupied mainly by bankers, money changers and jewellers, should accrue to them, a view not upheld by the courts, which agreed with the city that the monks' privileges pertained simply to fishing rights. Although the city had an obligation to use the rentals to keep the bridge in first-class condition, it did not do so. The worm-riddled members, having suffered two harsh winters in 1480 and 1497, when the Seine iced up, finally gave way in 1499 with a loud cracking noise, and the bridge collapsed into the river, taking a number of persons with it. The courts held a number of leading city officials guilty of corruption and handed out fines and gaol sentences as well as ordering them to reimburse the tenants of the houses for their losses.

Within two weeks of the collapse the decision had been taken to replace it with one of stone. As always, the financing of the new bridge presented a problem, but the city authorities solved this by imposing a levy on cattle, fish and salt. Louis XII appointed an Italian monk, Giovanni Giacondo, as designer and engineer in charge of construction of the new Ponte Notre Dame. He had been brought to Paris, together with other artists and engineers, by Louis XII's predecessor Charles VIII, whose appreciation of all things Italian stemmed from some swift conquests he made in that country in 1494 (encouraged by Leonardo da Vinci's Milanese patron Lodovico Sforza) in his efforts to assume the Kingdom of Naples. Within a year, with Ludovico having changed sides, he withdrew from Italy and died soon after. Louis XII closed the circle by capturing Milan in 1499 and driving out Ludovico.

Giacondo designed a bridge with six semi-circular arches to cross the 124 metre wide river, the central four arches having spans of 17.3 metres and the outer two being slightly smaller. Pier widths of 4.5 metres gave span to pier ratios of 3.8. A notable feature of the structure was the introduction of corne-de-vache arch openings, a technique copied by a number of later outstanding designers of stone arch bridges, including Perronet for the Neuilly Bridge over the Seine (1772) and Telford for the Over Bridge in Gloucester (1830). Corne-de-vache means cow's horn and describes a span in which the spandrel wall opening has a larger radius than the barrel of the arch, which gives a funnel-like flaring-out at the haunches of the arches. This serves both to give additional width to the roadway and to conduct flood waters smoothly through the arched opening. The Pont Notre Dame had a total width of 23 metres, but a double row of four-storey houses took up to 16 metres of this, leaving a street width of only about 7 metres, a central gutter serving to drain the roadway. It says much for Giacondo's organisational and administrative skills that he took only seven years from 1500 to 1507 to build the bridge.

Having adopted pier widths only slightly greater than a quarter of the arch spans, Giacondo must have felt that he had a fairly comfortable margin in avoiding scour around the base of the piers, but he had reckoned without the avarice of the city authorities, who allowed a weir to be built under one arch and mills under two others. It is a testimony to the quality of the bridge foundations that while the increased current velocities through the arched openings hindered navigation, the structure itself suffered no damage.

Pier foundations were constructed successfully in the dry, within cofferdams from which the water had been removed by horse-operated pumps. Oak piles were then driven at slightly greater than one metre centre to centre spacing, unusually wide in comparison to usual practice at the time, but conforming well with modern practice. Layers of rubble concrete laid on top of the piles consisted of large stones placed in a mortar of river sand and pozzuolanic cement, the latter obtained from the neighbourhood of Meaux and Mellun, and seemingly the first recorded use of this substance since Roman times. The masonry piers with triangular cutwaters upstream and downstream were built up from river-bed level on these rubble concrete bases.

By 1786 the pressure of traffic needing to use the bridge had become so great that the Municipality decided to remove all the houses from the bridge, allowing the roadway to be widened to 13 metres, with 4.8 metre wide footpaths on each side. In 1853, again to accommodate traffic requirements, the structure of the bridge was demolished to water level and rebuilt with narrower piers and elliptical arches with a smaller rise, thus allowing the road level to be lowered.

After completion of the bridge Giacondo returned to Italy and concerned himself with the problems of Venice. Realising that silt from the River Brenta filling the lagoons threatened the security and commercial supremacy of the city, he drew up plans to dig a canal to divert a large part of the Brenta flow into the lagoon of Choggia. The Council ultimately commissioned Alessio degli Aleardi to carry out this scheme, which had the added advantage of turning part of the Choggia lagoon into good dry land. In 1512, when a fire destroyed much of the Rialto district and threatened the existing timber bridge, Giacondo prepared plans for a complete new layout for the area, but the owners successfully resisted and simply replaced the previous buildings. Giacondo's proposal to replace the bridge with a stone structure also fell on deaf ears and another eighty years passed before Antonio da Ponte completed his masterpiece.

Bridges, more even than buildings, can command a remarkable degree of pride and affection in the hearts and minds of people. Many examples can be cited: the Anchi Bridge over the Chiao River in China, the Chain Bridge in Budapest, the Charles Bridge in Prague, Brooklyn Bridge and the Golden Gate Bridge in USA and the Sydney Harbour Bridge. Pont Neuf in Paris falls into this category. It has inspired innumerable verses and sayings. It is said that if a Frenchman asks a colleague, '*Comment vous portez-vous?*' he may well receive the reply, '*Fort Comme le Pont-Neuf*'! In fact, for much of its life the bridge was a sick patient, enfeebled by foundations weakened by scour and disintegrating voussoirs in some of the arches. It may well be that the great affection Parisians have always had for this bridge stems in some measure from the bridge's imperfections, which extend also to the rather curious geometry of some of the arches, with different span openings upstream and downstream.

With a rapidly growing population north and south of the Seine in the sixteenth century, a pressing need developed for a bridge to connect the two banks to the Ile de la Cité, which contained the Royal Palace, the Cathedral of Notre Dame and many public buildings. The conflicts between Protestants and Catholics, however, created a climate of uncertainty not conducive to undertaking a major work of construction in the capital. On 24 August 1572, matters came to a head when Catholic nobles persuaded Catherine de Medici, who had generally adopted a policy of reconciliation, to launch an attack against the Protestants on the pretext of their planning an assault on the Louvre. She, in turn, convinced her son Charles IX that the attack should be made; it became known as the St Bartholomew's Day Massacre, in which at least 2,000 people died. Charles IX died soon after, to be replaced by his brother Henry III, who signed a peace treaty with the Hugenots in 1577. With military distractions temporarily out of the way, Henry commanded that a new bridge be built and work began in 1578; however,

The Pont Neuf, completed in 1607, connects the two banks of the Seine in Paris to the Ille de Cité, the southern section of five spans having a total length of 80 metres and the northern end having seven spans totalling 150 metres. No two arches have the same span. It underwent extensive reconstruction in 1848.

the fragile peace did not last and violence flared again, with the Catholics incensed at Henry aligning himself with the Protestants. Work on the bridge stopped. In 1589 Henry III was assassinated by a Dominican monk, to be replaced by the much more politically astute Henry of Navarre, who became Henry IV. In 1593 he renounced Protestantism, thus securing the throne and assuring peace within the country. Work on the Pont Neuf suffered further delay, however, as Henry IV declared war on Spain in 1595, which lasted until 1598, when the two countries made peace. In 1599 work on the bridge resumed, leading to its completion in 1607.

When Henry III ordered the bridge to be built, he appointed a Commission to take charge of the work. In their early deliberations, they clearly realised that the provision of stable foundations posed a major problem, which prompted them to seek advice from a number of expert master masons and others on their construction. One of the experts, a consulting engineer named Pierre Lescot, advised that the ground under the south abutment or first pier should be thoroughly explored before deciding on which to adopt of the three foundation solutions in favour at the time: piles driven to refusal, a timber grillage placed on the base of the foundation excavation to support the pier, or stone ashlars placed directly on the base of the excavation. On the basis of results from the exploration made of ground conditions, and despite reservations expressed by Lescot at the rejection of piles, the Commission chose to use grillages comprised of heavy planks 300 mm thick, laid close together and resting on stringers or sleepers 750 mm thick. This solution proved inadequate owing to the highly erodible nature of the sands underlying the grillages, giving rise to severe scour. Ironically, these sands were only about 3 metres deep and thus easily penetrable by piles, which would have gained excellent bearing on underlying rock. Construction of each pier foundation took place inside a coffer dam consisting of an inner double row of timber sheet piling with puddle clay filling the intervening space, and a single outer row of piles protected on the outside by a sloping earthen embankment. When work stopped in 1588, a number of piers had been constructed and removal of sand below the grillages by scour had caused some of these to settle badly. Attempted repairs made within temporary cofferdams were followed by the construction of permanent cofferdams or starlings surrounding the piers; these comprised piles, driven to refusal, about 2 metres outside the perimeter of each pier, cut off below the water surface and anchored to the piers by iron ties, with the interior space then filled with rubble. When this solution proved unsuccessful, further strengthening attempts followed at intervals of time with varying success, relying either on the placing of rip-rap or the use of thicker piles for the protective cofferdams.

The bridge crosses the river on a 10° skew, the short southern end of five spans and the longer northern end of seven spans having total lengths of 82 metres and 150 metres respectively. Not only do all the arches have different span lengths, ranging from 9.8 metres to 15.8 metres in the southern arm and 14.7 metres to 19.4 metres in the northern arm, but most of them have different span lengths upstream and downstream, the fourth span from the bank, for example, having spans of 15.5 metres upstream and 13.5 metres downstream. All arches approximate to a semi-circle in shape. Some of the original poor workmanship resulted in arch rings thinner than prescribed and lacking a cement coating over the extrados as specified and this, together with deterioration of the voussoirs, led to crumbling of the arches. The burning in 1778 of a water mill located below the centre arch of the long arm caused additional damage to the span.

Extensive reconstruction in 1848 changed significantly the appearance of the bridge. The long arm of the original structure had ascending gradients from the ends to the middle of 3.4 per cent, which the ever increasing volume of wheeled traffic found difficult to negotiate. In order to eliminate the steep gradients, the original semi-circular arches, by now in a bad state of repair, were replaced by elliptical curves. The arches of the short arm had a corne-de vache geometry, having been widened during construction to enable shops to be built on both sides of the road; the arches were repaired but not reconstructed in 1848, as the road gradients over this length were not steep. The shops never materialised. Henry IV allowed only bookstalls and other small merchandise stalls to be set up on the bridge and they had to be taken down each day, but even these caused severe congestion on the 4.5 metre wide footways. With footway levels two steps above the level of the 11.2 metre wide roadway, pedestrians needed to exercise care at all times, although circular balconies over the piers provided some refuge.

From the completion of its construction, the bridge attracted a wide and motley collection of people, described vividly by a seventeenth-century observer:

> The bustling activity of the middle class, the awakened loitering of the gaping loafer, the rollicking boisterousness of the insignificant cadet, the poverty of the monks, the

insolence of the public women, the arrogance of the noble lords as they stroll in gallant
company, the forwardness of courtesans as they move towards the Louvre, horsemen,
pedestrians, carriages, sedan-chairs, all rolled on in a never-ending stream. Charlatans,
fortune-tellers, sellers of nostrums, and montebanks there selected their stands and
there attracted idlers, thieves, swordsmen, cloak-snatchers and pickpockets.

As with any major bridge, it has witnessed many events of history. In 1610 the body of
the assassinated Henry IV crossed the bridge to its final resting place in the crypt of St
Denis in Notre-Dame, and seven years later his widow Marie de Medici crossed it to exile
in the provinces. Her replacement of Sully by the Italian Concini roused the anger of the
Parisians and Henry's son Louis XIII had Concini shot, and the citizens hung his body
from a gibbet on the bridge. A much happier occasion in 1660 saw the Spanish Princess
Marie-Thèrése, bride of the new King Louis XIV, cross the bridge on her way to Notre-
Dame, followed by a large retinue.

Towards the end of the eighteenth century the revolution swirled around and across
the bridge: the tumbrels carrying royalists and revolutionaries to the guillotine passed the
end of the bridge but crossed the river by the Pont-du Change, after which they turned
left down the Quai de la Mégisserie towards the Place de la Révolution. A departure from
this route was made in the case of the revolutionaries Robespierre, Coutlon and Saint-
Just, who were taken across the Pont Neuf to the guillotine because it was sufficiently
wide to be lined by troops.

Until the eighteenth century, one characteristic linked all multi-span masonry arch
bridges – the use of wide, heavy piers that blocked substantial portions of the river
flow. The span to pier ratio commonly fell within the range of 3–5, whether Chinese,
Persian, Roman, medieval or Renaissance. Protective measures such as starlings, taken
to protect the piers, often reduced the flow area even further. The resulting rapid flow
of water through the openings often posed a constant threat to the stability of the
bridge foundations, but in some instances proved a boon to mill owners who used the
flow to drive their waterwheels. Each pier, in effect, became an abutment, which gave
an advantage during construction as each arch could be completed independently of
the others, and also destruction of an arch during military operation did not lead to
collapse of the whole bridge. The final act in the perfecting of multi-arch masonry bridge
construction came in the eighteenth century in Paris in the designs of Jean-Rodolphe
Perronet (1708–94), but it came at a time when masonry bridge construction was about
to give way to the new materials – iron, steel and concrete.

His appointment in 1750 as Chief Inspector, and later Chief Engineer, of the Corps
des Ingénieurs des Ponts et Chausées, allowed him to introduce several important
innovations into bridge building. Concerned about the restriction to water flow (and
hence possible scour), and also to traffic using the river, caused by the conventional use
of wide piers for multi-span masonry arch bridges, he implemented the idea of carrying
horizontal thrusts from the arches through to the abutments, thus requiring the piers to
take only the vertical loadings of the bridge and bridge traffic. This enabled him to use
flatter arches and much reduced pier widths, but required all spans to be approximately
equal to ensure against unbalanced thrusts at the piers, and additionally required a

modified construction procedure of removing centring from all arches simultaneously. He also adopted the corne-de-vache geometry for his arch openings to facilitate the passage of flood waters.

His first bridge built on these principles over the Seine at Neuilly, just north of Paris, and completed in 1772 consisted of five flat segmental arches of 37 metres span, with pier widths of only 4 metres, little more than one-tenth of the span. The arches had a rise to half-span ratio of 0.5. His most daring bridge, over the Oise at Ste Maxence just north of Paris, comprised three arches of 23.5 metres span each, each with an almost unbelievable rise to half-span ratio of 0.19 and a span to pier width ratio of 8.5. As a further innovation, he adopted double-column piers. His final masterpiece, the Pont de la Concorde in Paris, completed in 1791 when Perronet was 83 years of age, differed slightly from his own conception, as he was forced by pressure of criticism to increase the rise to half-span ratio of the arches. He managed, however, to resist demands to thicken the piers and countered the unwelcome modification to his arch geometry by providing a balustrade to emphasise the lightness of the bridge, and it remains a classic to this day.

7

Castles and Fortresses

Byzantine

Beset by Huns and Goths to the north and Persians to the east, the Byzantine emperor Theodosius II commissioned his prefect Anthemius in 413 AD to build land walls to surround Constantinople. The resulting wall, 5.6 km long, 4.5 metres thick and nearly 12 metres high, had the added protection of ninety-six projecting battlemented towers up to 18.9 metres high, some square and some octagonal in shape. In 447 AD, a severe earthquake caused extensive damage to the wall at the same time as Attila the Hun threatened dangerously from the north, causing Theodosius not only to order restoration and strengthening of the walls but to add a second lower outer wall 2 metres in thickness with ninety-two towers located at intermediate points between the towers of the inner wall. A moat 15.2 metres wide and 5.7 metres deep provided further protection. Constructed with a concrete core, faced with limestone blocks bonded at intervals of height by horizontal brick-lacing courses, five bricks in thickness, a pattern that became common in Turkish architecture, the walls protected the city for a thousand years until breached by the cannons of the Ottoman Sultan, Mehmet, in 1453.

The pattern of walls built to protect Nicaea in Asia Minor, also constructed in the fifth century, although added to later, corresponded closely to the double wall system at Constantinople, with an inner wall much thicker and higher than the outer wall. Brick lacing courses were again incorporated into the stonework, with numerous towers set into the walls spaced so that each tower on one wall was set opposite an interval between towers in the other wall. Four main gates and three minor entrances, posterns, gave access to the city. After falling into the hands of the Sultan these fortifications, which are still in a good state of repair, repelled repeated attacks by the Crusaders until undermined by these assailants in 1097.

Faced with the need to protect his extensive empire, the Byzantine emperor Justinian not only strengthened existing fortresses and built numerous new ones but introduced new concepts that influenced later fortress builders, both Christian and Muslim. At Dara and Theodosiopolis, near the Persian frontier of his empire, he had the existing walls heightened by constructing a gallery along the tops of the walls surmounted by a second line of battlements. He also had a second curtain wall built outside the first, leaving a gap

The Theodosian walls protected Istanbul for 1,000 years until they were breached by Ottoman cannon fire in 1453.

of 15 metres between them. In Dara, as in all his defence works, Justinian paid particular attention to ensuring a protected and adequate water supply. A stream emerging from the mountains, which also provided protection, was diverted to enter the walls through iron-barred conduits, to flow through the city before passing out on the opposite side, again through iron bars set in the walls. In times of siege, the water could be diverted into an underground passage with an outlet many miles away to deprive the assailants of any supply.

Early and Medieval Europe

In northern Europe, following the demise of Rome's military authority, reliance for physical protection rested very largely on the remains of Roman fortifications, particularly fortified towns and cities, and even, in some cases, earlier Iron Age fortifications. When the need to construct new defences eventually became unavoidable, they initially took the form of earth or timber structures, or, commonly, a combination of these. The earliest in northern Europe, the gród, featured a ring fortification consisting of an earthen rampart and moat, wooden palisades and a fortified gate. It differed little from fortified sites dating back for a thousand years.

Earth and earthworks formed a major part of the motte and bailey castles introduced into Britain and Normandy by the Normans, examples of which occur also in Germany, Italy and Denmark. As the last line of defence the motte, supporting a timber donjon or masonry keep, comprised a mound dominating the site, sometimes consisting entirely of artificially placed earthen materials, but on occasion incorporating raised topographical features where such existed, thus reducing the amount of material and effort required to form the mound. Utilisation of natural features such as a promontory neck, cliff face or ravine could measurably reduce the amount of earthworks needed to protect a site.

Known examples of mottes range from 3 metres to 30 metres in height with base diameters of 30 metres to nearly 90 metres. The flattened tops of the mounds ranged from 10 metres across to 40 metres or more. A scene on the Bayeux Tapestry shows the mound of William's castle at Hastings in the course of construction, made up of rammed earth and stones. Open areas within the protected enclosure became known as the bailey, a term originally applied to the enclosure wall and is enshrined in the name of the Central Criminal Court in London, the Old Bailey.

With the development in construction technology, and in weapon technology, came greater sophistication in the building of fortifications, notably in castle construction, with masonry largely replacing earth and timber. These masonry structures themselves evolved over time from high walls to protect against bows and siege engines to complex massive masonry structures and ramparts needed to protect against cannon fire.

The moat, a feature common to most medieval castles and usually 3 metres or more in depth, not only constituted a defensive element itself, but the excavated material provided all or much of the earth forming the ramparts within the moated area. Although often left dry, unless excavated to depths below the local water table, moats still presented a serious obstacle to siege engines and assaults on the walls. Where a fortified position occupied the crest of a hill, with a circling moat excavated into the hillsides surrounding it, added protection could be gained by throwing the excavated soil downhill of the excavation rather than within the protected area, a method known as scarping, which gave steepened faces, difficult to scale, both downhill of the moat as well as on its excavated inner face.

With the passage of time and the development of more effective assault weapons, the timber donjon gave way to a masonry structure more commonly known as a keep. Timber palisades used for early enclosure walls in turn gave way to much more substantial stone outer walls and gatehouses. As the masonry keeps became more substantial, the artificial mottes became increasingly irrelevant and in any case could not support their weight, and the massive keep displaced the motte as the strongpoint of the castle. The square shape favoured for early masonry keeps tended to give way to a circular form by the end of the twelfth century.

The medieval period in Europe saw a spate of castle construction, with massive keeps as the last line of defence and high battlemented curtain walls and towers to keep defenders out of the reach of arrows and missiles, which they could in safety rain down on their attackers. Corbelled or cantilevered projections at the tops of the walls allowed defenders to pour down boiling water and combustibles onto their assailants. High walls and towers also gave the defenders elevated positions from which to scan the surrounding countryside for possible danger. Towers usually projected beyond the outer faces of the walls so that flanking fire could be directed along the walls. In order to be fully effective, the towers had to be sited no more than one bowshot apart so that each tower could protect its immediate neighbours.

In the five years following the Battle of Hastings, William built thirty-three castles as he brought the whole of England under his control by a series of military campaigns and marches, and in the remaining sixteen years of his reign up to his death in 1087 he almost tripled this number. But once the Normans had seen off the danger of Saxon uprisings and Danish invasions of England, the nobles proceeded to fight both among themselves and with the King, leading to a burgeoning of castle construction towards the end of the eleventh century. A further spur to castle building came with the civil war (1135–53) between Matilda, Henry I's only legitimate surviving child, and Stephen, Henry's nephew, who had been crowned by a group of influential barons. Finally in 1153, both sides, now tired of conflict and fearful that the French might exploit the state of anarchy in England to capture lands in Normandy, signed a peace treaty allowing Stephen to remain as King for the remainder of his life, to be succeeded on his death by Henry of Anjou, Matilda's son. And so, on Stephen's death in 1154, there came to the throne one of the most formidable of British monarchs and one of the great castle builders, a role forced upon him to nullify the advantage in castle ownership that lay with the barons at the time of his succession. Henry II spent nearly a tenth of the annual revenues of the crown on building the great stone castles. The money disappeared into the cavernous demands of quarrying, dressing and transporting stone, producing lime and mortar, and paying the wages of masons, carpenters and other craftsmen, as well as the masses of labourers who hacked out the soil and rock for the foundations and ditches.

He commissioned 'Maurice the Engineer' to convert the Conqueror's castle on its prominent site at Dover into the strongest fortress in the land by constructing a massive stone keep almost 30 metres square and of similar height, with buttressed walls up to 6.5 metres thick. A cross-wall bisects the building and divides each floor into two long halls. It became the first castle in western Europe to adopt the Theodosian system of two independent concentric walls, with the height of the inner wall exceeding that of the outer wall to give its defenders protective fire. Fourteen rectangular towers are set into the wall of the inner bailey, four of them paired to give two gatehouses, unique in Britain at the time of their construction. Projecting towers in the outer wall gave it protecting flanking fire. Henry II spent £7,000 on Dover castle at a time when the total royal revenues amounted to £10,000 per annum.

Edward I, considered one of England's greatest kings, devoted his life and considerable energy and abilities to creating a United Kingdom of England, Wales and Scotland. He also possessed a genuine wish to bring law, order and prosperity to the land, and towards accomplishing this set up the Model Parliament in 1295 to provide a system of representative government, which has evolved over the centuries to become the parliamentary system in the UK today. In order to achieve his aim of a United Kingdom, he had to subjugate both Wales and Scotland and turned his attention in the first place to Wales. While the terrain in the south of the country consisted of valleys and rivers that presented no great difficulties to invaders, the north presented a maze of steep mountainous country, deep valleys, some with marshy bottoms, and very wet, often foggy, weather: ideal conditions for harrying invaders and quickly disappearing into the hills.

Well aware of the difficulties facing him, Edward decided to overcome, by sheer weight of numbers, the natural hazards and his powerful adversary Llewellyn, who had successfully united north Wales behind his cause. Llewellyn ap Gruffydd had also made overtures to the French for support, to the extent of taking for his bride a lady of the French court. Unfortunately for both of them, she fell into Edward's hands on the way to meet her intended groom. Llewellyn's defiance crumbled in the face of the vast invading army that had him trapped with nowhere to go, and after swearing allegiance to Edward he was united with his bride. But, although Edward treated him well, the Welshman was biding his time and in 1282, together with his brother Dafydd ap Gruffydd, attacked, burnt or killed everything English or every English person they could lay hands on. Edward again assembled a powerful army against which Llewellyn quickly succumbed, to be banished forever from Wales, but he lost his life in an apparently minor skirmish or entrapment shortly after. Edward subdivided Llewellyn's Principality of Gwynedd into three districts, Anglesey, Merioneth and Caernarfon, and set about building a system of roads and castles so that every important area of the country could be readily accessed and defended. Caernarfon and Harlech are just two of the many castles he had built or substantially rebuilt, most of them within the space of little more than a decade.

Harlech castle, built over a period of nearly seven years, 1283–9, by Edward's master mason James of St George, occupies a spectacular cliff-top site, its construction making full use of the topography, not only in exploiting its defensive characteristics but also, as long as the occupants had command of the seas, in ensuring supplies to the occupants during times of siege. At the time of construction the sea, now kilometres away, lapped the foot of the cliffs and the castle could be accessed from the sea by a protected stairway cut into the rock and running 61 metres up the cliff to enter the fortress through a fortified gate.

Rectangular in plan, the inner bailey or court is protected by massive inner walls up to 24 metres high and up to 3.5 metres thick, with additional protection in the form of much shorter and less substantial, now largely ruined, outer walls separated from the inner walls by a narrow passage or middle bailey, which is interrupted at one point by a cross-wall and doorway. A wide moat, excavated deep into the rock, protects the east and south sides and an outer bailey extends down the steep slope on the north and west sides. The inner walls have a large round tower at each corner and a massive gatehouse flanked by D-shaped towers in the middle of the east wall. Two posterns in the north and west walls, the latter leading directly out of the great hall, give access to the middle bailey, from where the outer bailey can be accessed by two posterns in the outer walls. The outer walls have no towers and a small gatehouse, with a two-leafed door, accessed across the moat by a drawbridge and itself giving access by a short passage to the massive inner gatehouse protected by a stout timber bar, three portcullises and two doors, and from above by eight wide machicolations.

Built at a cost in excess of £8,000, the construction work force at its peak numbered around 820, comprising some 230 stonemasons, 20 carpenters, 30 blacksmiths and the remainder labourers. In addition, over 100 were employed in the quarries. On its completion Edward appointed James to the position of Constable of Harlech Castle, from where, over the next three years, he continued to supervise the construction of Edward's other strategically located castles fringing the coast of Snowdonia.

The castle proved its worth in 1294 when its occupants, able to be supplied from the sea, successfully held off a siege by Madoc ap Llewellyn, who had already razed a number of English-held towns. In 1404, however, after a long siege had reduced the garrison to only twenty-one occupants, the castle fell to Owain Glyndŵr, who had been proclaimed Prince of Wales by his rebellious band of followers in 1400, many of whom had previously seen English service as archers or men at arms against Scotland and France. Using it as his residence and military headquarters, he also set up his parliament here in 1405; however, his short-lived occupancy lasted only until 1409, when the English retook the castle, capturing Owain's wife and two of his children and incarcerating them in the Tower of London, where they all died within a few years. Owain himself escaped, only to be captured in 1412 and ransomed, subsequently disappearing without trace. During the War of the Roses the castle became a Lancastrian stronghold and was in 1468 the last fortress to fall to the Yorkists, having been able to endure a seven-year siege through being provisioned from the sea. According to tradition, the song 'Men of Harlech' received its inspiration from this siege.

Unlike Harlech Castle, Caernarfon Castle, also the work of Edward I's master-mason James, was built on essentially level terrain on the banks of the River Seiont where it flows into the Menai Straits. A Roman fort and a Norman motte and bailey castle had previously occupied the site, emphasising its strategic value. Shaped like an elongated hour-glass extending some 160 metres east–west, its north wall faces towards, and practically abuts, the town of Caernarfon, also built by Edward to replace the existing Welsh town, and the castle is directly linked to, and in effect lies within, the town walls. The south wall faces onto the river. Moats protect the northern and eastern sides of the castle. The high walls, up to 6 metres thick and still in a good state of preservation, are studded with nine polygonal towers, most having high observation turrets, and two gatehouses. The unusual shape and plethora of towers enabled the defenders to launch arrows and missiles from many different angles and from various heights. The King's Gate, situated in the centre of the north wall, provided direct access from the town into the western or inner bailey of the castle, which housed the offices of state, accommodation quarters and the Great Hall, of which only the foundations now remain. Again, unlike Harlech, which was constructed solely for military purposes, Caernarfon was also intended as a seat of government, in effect a palace, as well as a military fortress. Edward's son, later to be the incompetent King Edward II, was born here and invested with the title 'Prince of Wales', the castle being similarly used in 1911 for the investiture of the future Edward VII as Prince of Wales, and again in 1969 for the investiture of Prince Charles. The smaller Queen's Gate at the east end of the castle, built on top of the original Norman motte and reached by a stone ramp from ground level, provided access from outside the town walls into the eastern or outer bailey. The most prominent tower, the Eagle Tower, virtually a keep in itself with its own private chapel, and situated at the extreme western end of the castle, links directly into the town walls and is a dominant feature of the whole fortress. It had its own postern gate, the Water Gate, opening onto the river, allowing private access or quick escape if necessary. Its name apparently derived from carvings of eagles ordered by Edward for the tower, reflecting his veneration for the military prowess of Imperial Rome. Other echoes of Roman influence are seen in the banded patterns of the masonry towers, a feature copied from the fortified walls of Constantinople, which Edward no doubt remembered from his time as a young man travelling in the Holy Land with his new Spanish wife.

Construction, which commenced in 1283, consisted in the first place of demolishing the old town and rebuilding it, excavating the moat separating the town from the castle, erecting wooden barricades followed by the Eagle Tower as a holding defence strategy, then building the remaining walls, other than the north wall, but including the walls linking up to the town walls. This reliance on the moat and town walls to protect the northern side proved to be no deterrent to Madog ap Llewellyn, a cousin of Llewellyn ap Gruffyd, who in his revolt against the English in 1294 successfully attacked through the town and crossed the moat to take the castle. The English retook the castle the next year, and in addition to repairing the damaged walls immediately set about constructing the north wall and its massive King's Gate.

Despite taking some eighteen years in the building, much longer than any other of Edward's Welsh castles, some parts were never completed as originally envisaged, most notably the King's Gate, planned originally to extend across the width of the castle to the south wall with five doors, six portcullises and divers spy holes, arrow loops and murder holes. Nevertheless, this massive twin turreted gate as constructed still had formidable defence capability, particularly in conjunction with the town walls which, themselves, were surrounded by water-filled moats, the rivers Seiont and Cadnant and the Menai Straits. All materials for building the town and castle had to be brought in by sea, requiring the construction of a special quay, adding to the cost of about £12,000

Carnarfon Castle, built in 1283 by Edward I's master mason James, occupied a strategic position dominating the River Seiont and the Menai Straits in North Wales. A Roman fort and a Norman motte and bailey castle had previously occupied the site.

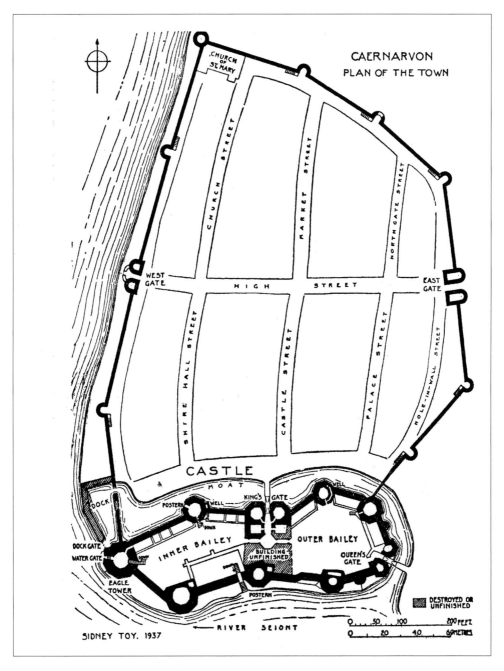

CAERNARVON
PLAN OF THE TOWN

CHURCH OF ST MARY

CHURCH STREET

MARKET STREET

NORTH GATE STREET

WEST GATE

HIGH   STREET

EAST GATE

SHIRE HALL STREET

CASTLE STREET

PALACE STREET

HOLE-IN-WALL STREET

CASTLE

MOAT

WELL

KING'S GATE

DOCK

POSTERN   WELL

DOWN

DOCK GATE

WATER GATE

DOWN

INNER BAILEY

BUILDING UNFINISHED

OUTER BAILEY

QUEEN'S GATE

EAGLE TOWER

POSTERN

DESTROYED OR UNFINISHED

0 50 100   200 FEET
0   20   40   60 METRES

SIDNEY TOY. 1937

← RIVER   SEIONT

The castle links directly into the walls of Carnarfon town.

in the initial period up to 1292 and a final cost of around £22,000, somewhat in excess of the royal treasury's annual income. It did, however, earn its keep in 1403 and 1404, when it withstood sieges by Owain Glyndŵr and, unlike Harlech, remained in English hands. During the Civil War, its Royalist occupants surrendered to Parliamentary forces in 1646.

Mediterranean

One of the most imposing castles of the medieval period built by Westerners was not in Europe, but in Syria. Having inherited the already fortified site of Le Krak des Chevaliers in 1142, the Knights Hospitaller built most of the structures now on the site, with the exception of a large square tower in the outer wall built by the Moslems in 1285, the citadel having fallen to Baybars in1171. Occupying a formidable site with sheer drops on the east, west and north sides and with a moat on the more gradually descending south side, its massive double walls with towers and battlements made it virtually impregnable. An outward splay at the base of the outer walls increased their resistance to mining and tunnelling, as Baybars found when he lay siege to the castle in 1171. After one month he did manage to break through the outer walls, but could not then penetrate the inner walls. But where force failed, simple guile succeeded. He had a forged letter, ostensibly from a friend, delivered to the Grand Master advising him that Tripoli had fallen, rendering further resistance pointless; the garrison troops, by now in a state of exhaustion, were ready to believe anything and surrendered the castle.

Founded by the Blessed Gerard after the First Crusade to care for pilgrims in Jerusalem, the Knights Hospitaller, while continuing to work for the sick, also acquired a powerful military status comparable to that of the Knights Templar. At their peak strength in the Kingdom of Jerusalem, they held another six major forts in addition to Le Krak des Chevaliers, and many other estates in the area; following the fall of Jerusalem itself in 1187, they continued to lose out militarily to the Muslims until Acre became their

The Krak des Chevalliers was one of a number of major forts built in Syria in the twelfth century by the Knights Hospitaller.

sole possession, and when this city fell in 1291 they sought refuge briefly in Cyprus, at that time the possession of the French Lusignan family, who had acquired it from Richard Lion-Heart; here, under Grand Master Guillaume de Villaret, the Hospitallers, now the Knights of St John, set about planning a more permanent home on the island of Rhodes, eventually retrieving it from the Byzantines in 1309. They set about greatly strengthening the fortification of the city of Rhodes itself, occupying a site on the east shore of the northern tip of the island. Close to the site where the Colossus of Rhodes once stood, two well sheltered harbours were formed, flanked and protected by three moles running out from the shore, each having a powerful masonry tower at its sea end. The most northerly of these, the St Nicholas tower, a strong circular structure with powerful bulwarks, storerooms and a well, overlooked the ancient harbour from its western edge and had the potential to withstand a long siege. An iron chain stretching between the other two towers, St Angelo's and Naillac's, at the entrance to the Grand Harbour provided additional protection from a possible seaborne attack.

The walls surrounding the town were initially constructed with thicknesses from 3 metres to 4 metres and 9 metres high, with crenellated parapets, strengthened at intervals with towers, and a moat on the landward sides. Realising the inadequacy of these walls against artillery, Grand Master Pierre d'Aubusson (1470–1503) put in hand a programme in 1470 to have them substantially strengthened. The work entailed for the most part widening and deepening the moat, construction along some lengths of an outer rampart and second moat, and bulwarks around gateways and salient towers on which heavy guns could be mounted. Three of the original eight gates were blocked to increase security, and others were strengthened. One gate, the Koskino Gate, in addition to the low-level bulwark had a short intermediate moat separating the bulwark from a bastion fronting the drawbridge over the inner moat, which had two more levels on which heavy guns could be mounted. An outer moat, inner stout door and portcullis reached by a zig-zag passage completed its protection. Other strengthening measures included placing additional material on the inside of the walls, increasing their thickness to about 12 metres, modifying the tops of walls to maximise the employment of defensive artillery, and lowering the heights of towers to curtain wall level.

The first great siege by the Muslims came in 1480, when an amphibious task force of 160 ships under Meshid Pasha attacked the city by order of Sultan Mohammed II, who had failed to persuade Pierre d'Aubusson that the Knights should pay him an annual levy. The Knights were divided into eight sections according to their native language – Aragon, Auvergne, Castile, England, France, Germany, Italy and Provence – each manning a designated portion of the walls. Their initial assault on the west walls having been repulsed, after succeeding in landing on the west shore of the island, the attackers decided to concentrate their offensive on the tower of St Nicholas on the advice of a German engineer in their pay (who was later hanged in the city square as a spy, having sought the forgiveness of the Knights). The Turks bombarded the tower with heavy guns mounted on the eastern shore of the Old Harbour, but their efforts to bombard it from the sea were largely thwarted by the Knights having stationed fire ships in the harbour and fixed spiked planks in the sea bed at its entrance at the entrance. Their heavy artillery and Greek fire wreaked havoc on the fleet, destroying 700 ships.

Having merely damaged the tower but failed to capture it, the Turks once again turned their attention to assailing the walls of the city, soon breaching some sections with their heavy artillery firing huge cannon balls. In a surprise night-time operation, Pierre D'Aubusson sent out fifty men, who succeeded in putting many of the Turks to the sword, destroying their siege positions and spiking their guns. With this defeat, the Pasha again turned his attention the St Nicholas tower, his engineers devising the extraordinary plan of building a timber bridge on land and then, by means of a cable passing through a ring anchored to the rock below the tower, towing the bridge into position to span the entrance to the Old Harbour, allowing access to the tower from the west shore. An Englishman named Roger swam underwater and released the anchor (for which he received 200 crowns of gold), but the besiegers eventually manoeuvred the bridge into place using floating craft. Furious fighting ensued, the Turks attacking across the bridge and from the sea only to see their fleet further devastated by cannon fire from the tower and the walls, which also smashed the bridge to pieces, killing all those on it.

The Turks turned their attention once again to attacking the walls at various points and in the process began raising an access ramp of earth to reach the full height of the wall. Seeing the danger, D'Aubusson directed a huge trebuchet at the works, demolishing them with its massive stone missiles and killing the workers. After desperate fighting on both sides, the defenders turned the battle, became the attackers and routed the enemy. In a last minute desperate ploy, the Pasha deputed a small group of picked men to try and reach the Grand Master and kill him. They failed in this but managed to wound him before being overcome. After a siege of three months, what remained of the fleet left the island.

The Ottomans returned in 1522, now under the direct command of the Sultan himself, Süleyman the Magnificent. His huge army of 200,000 men mounted seven savage assaults on the walls, which had been repaired and strengthened following the 1480 siege and damages caused by an earthquake in 1481. His attacks having been repulsed with huge losses of men, disease ravaging his army, winter approaching and campaigns in the Balkans on his mind, Süleyman proposed, after six months, an honourable armistice, on condition that the embattled Knights of St John, faced with inevitable defeat, depart from Rhodes. They left for Sicily on 21 December 1522, but in 1530 reluctantly accepted an offer made in 1524 from Charles V of Spain, King of Sicily, to re-establish themselves permanently in Malta, for an annual fee of a single Maltese falcon payable to the Viceroy of Sicily. This fee also included the fortress of Tripoli in North Africa. The Knights' reluctance stemmed from the fact that the islands of Malta and Gozo had few natural resources, requiring items like foodstuffs and timber to be imported from Sicily, and existing Norman fortifications were totally inadequate for protection against the Ottomans. They chose to base their headquarters on an inlet, on the north coast of Malta, which was divided into two by the Sceberras Peninsula, projecting into it in a north-east direction, with the Marsamuscetto Basin to the north of it and the Grand Harbour to the south. The Sceberrus Peninsula, a rugged limestone outcrop with steep cliffs, was eventually chosen as the site on which to build the town of Valetta.

The Grand Harbour itself had two smaller parallel peninsulas, Birgu and Senglea, projecting into it from its southern shore and the Knights, with limited resources at their disposal, chose to upgrade the existing defensive structures on these two tongues of land rather than building a new, stronger fortress on the Sceberras Peninsula, despite its more prominent position. Fort St Angelo, a squat masonry structure with provision for artillery positions, was built on Birgu and massive masonry walls constructed to protect the outer shorelines of the two peninsulas. Instead of walling the inner facing shorelines, a chain was slung between the land tips to prevent enemy ships accessing the harbour area between the two peninsulas.

The first Turkish attack came in 1551 when an element of their fleet sailed into Marsamuscetto and disembarked several thousand men. However, instead of taking on the Knights' defensive structure, they laid waste to the northern part of Malta and to Gozo, which they practically depopulated in carrying off some 5,000 of its inhabitants into slavery. They then sailed on to Tripoli and captured its poorly defended fortress. This attack served as a wake-up call for the Knights and they put in hand the construction of the powerful star-shaped fortress of St Elmo at the tip of the still uninhabited Sceberras Peninsula, with modern outworks and artillery positions.

In 1565 Süleyman, now an old man with only a year to live after ruling for 46 years, sent a force of some 36,000 men to attack Malta and expel the Knights. There followed one of the great sieges of history, much written about but mainly by Western writers, so the reliability of the accounts is questionable. Grand Master Jean Parisot de la Valette apparently allowed the fleet to sail into Marsamuscetto without opposition and land the troops on its shores. Taking possession of the Sceberras Peninsula, they threw up an earth and stone wall along its spine to protect against artillery fire from the St Angelo fort on the opposite side of the Grand Harbour and excavated forward trenches to mount an attack on the St Elmo fort. In a surprise move, the fort sent out a detachment to attack the men in the trenches, but they were quickly overcome, allowing the Turks to occupy the outworks of the fort, which then came under artillery bombardment from a promontory called Gallows Point (now Ricasoli Point) on the opposite side of the Grand Harbour at its entrance, and from a promontory on the opposite shore of the Marsamuscetto at its entrance, called Dragut Point after the Ottoman corsair captain Dragut Reis, who organised the assault. After being beaten back numerous times in their attempts to scale the walls, crumbling under their relentless artillery bombardment, the Turks eventually overran the fort with the loss of 8,000 men, according to the chroniclers. The Christians lost almost their entire complement of 1,500 men manning the fort, including 130 knights.

The Turks now turned their attention to the Birgu-Senglea defences, which had been strengthened during the one-month siege of the St Elmo fort, initiating both land and sea attacks but themselves coming under a heavy barrage from the St Angelo fort, its guns on one occasion sinking nine of ten vessels containing 1,000 janissaries sent to try and gain a foothold on the outer defences. On another occasion, their troops streamed through a breach in the walls, only to be stopped by a hastily built second inner wall and now, trapped between the two walls, they perished under boiling water and pitch and other inflammable substances. With attack after attack being thrown back by the defenders, comprised largely of Maltese farmers and townspeople under the leadership of the knights, and beset by illness and shortage of food and ammunition, the

Turkish forces became increasingly dispirited, and, learning of the arrival from Sicily of reinforcements for the defenders, they broke off the four-month long siege and returned to Constantinople.

Turning his attention again to Hungary, Süleyman himself led a fresh campaign in 1566 to attack a fortress near Drava, but he died in his tent of natural causes shortly before the attack, his Grand Vizier withholding the news of his death from the army until they had achieved their success.

Following the departure of the Ottomans, Grand Master de la Valette put in hand the building of his new town of Valetta on the Sceberras ridge and employed the Papal engineer Francesco Laparelli to design a line of four bastions to enclose the landward end of the peninsula. With these fortifications complete, he was able to move his headquarters from Mdina into the new town in 1571. Major building projects continued in Valetta until the eighteenth century, including in the early seventeenth century the building of a much needed aqueduct to bring water to the city from the hills near Mdina. Much of the Order's income came from policing the seas, sometimes verging on piracy as they plundered ships with even the remotest Turkish connections, with little opposition after the destruction of the Ottoman fleet by the Holy Christian League in 1571 at the battle of Lepanto off western Greece, an operation in which the galleys of the Knights of St John took part. Among other things, the sea-going activities supplied a thriving slave market in Valetta. The Hospitallers' occupation of Malta came to an end in 1798, when the Order surrendered to Napoleon on his way to Egypt.

Renaissance and Later Europe

Towards the end of the fifteenth century, the advent of gunpowder and cannon in Europe started to have an influence on fortress design. The high vertical walls of the medieval castles proved particularly susceptible to the impact of shot and gave way to lower, thicker walls, presenting sloping or partly sloping surfaces to the enemy to deflect shot. Towers also had sloping surfaces at the base and circular and rectangular towers tended to give way to triangular and pentagonal shapes. The high, relatively thin walls of the medieval castles had the added disadvantages that narrow battlements could not accommodate the defenders' artillery, and firing from these heights caused the cannon balls to bounce harmlessly rather than scything into the midst of the attackers. Other innovations included star-shaped or polygonal fortresses, massive earthworks and wide ditches that did not become blocked when walls were breached, thus allowing easy access for the besiegers.

Not surprisingly, in view of the interminable struggles between the rich city-states, Italian engineers achieved a pre-eminence in fortress design in the fifteenth and sixteenth centuries. Leonardo devoted much of his thinking to military engineering and served briefly in 1502/3 as engineer-general to Cesare Borgia, the illegitimate son of Pope Alexander VI. His sketches reveal his own ideas both with respect to artillery and other devices for besieging and destroying a stronghold and to established designs that reflected general military thinking at the time. Drawings in the Codice Atlantico, which he made towards the end of the fifteenth century, show the development of the ravelin, a triangular projection usually placed in the ditch in front of the fortress curtain wall and giving flanking fire. It was, itself, protected by flanking fire from adjacent towers. Among other designs, Leonardo shows a square fortress with each side consisting of a tier of low-rising gun platforms set back from each other on continuous perimeter platforms, each protected by splayed walls. Leonardo also advocated hanging bales of saturated grass or hay on the front of defensive walls to absorb the impact of shot.

Michelangelo, given the exalted title of Commissar-General of Fortifications by the republican government of Florence, implemented rather crude, but apparently effective, measures to defend the city against the Medici supporters in 1530. He threw up earthen ramparts reinforced with timbers, fascines and crude bricks, which proved so effective in absorbing shot that the besiegers made no attempt to storm them, resorting instead to starving the city into submission over a period of nine months.

Francisco di Giorgio Martini, born in Siena in 1439, became at an early age municipal engineer in his hometown, responsible, among other things, for its water supply; however, in 1477 he entered the service of Duke Frederico da Montefeltro at the brilliant court of Urbino, specialising in military engineering, and is credited in this role with having built up to seventy fortresses, as well as designing various military machinery and weapons. After ten years in the service of the Duke he returned to Siena, responsible for both military and civil works as well as being in demand as a consultant to other towns and courts. Although he is believed to have built at least parts of the Palace of Urbino, his main charge lay in surrounding the town with a series of hill-top fortresses to protect its approaches. His designs for these retained some of the features of medieval fortresses such as machicolated walls, crenellated towers and a keep, while other features such as their polygonal shapes in plan, ranging from triangular to pentagonal, and even rhomboid, pointed to the future. Wedge-shaped towers incorporated into his designs were forerunners of the bastion system. The construction of the fortress of Ostia, designed by Giuliano da Sangallo and commissioned by Cardinal della Rovere, titular head of the Episcopal see of Ostia, to protect the entrance to the Tiber River echoed some of the designs of Francesco di Giorgio, its triangular plan featuring circular bastions at the three corner extremities and the preservation of an older polygonal medieval tower, acting as a keep, at the most acute corner, facing the direction of greatest peril.

An important development introduced by Italian military engineers towards the end of the sixteenth century was the bastion, projecting outwards at salient corner points from the walls of the fortress. The design consisted essentially of two sideways facing walls or flanking walls built out from the surrounding walls of the fortress and two closing faces, giving a triangulated aspect to the enemy. Cannon could be mounted on ramparts behind the walls to give outwards fire against the enemy, and also flanking fire along the ditches. Each bastion could also cover its two neighbours. The long, straight parapets of the bastions could also accommodate large numbers of cannon to direct onto the enemy. Credit for promoting this type of defensive structure probably belongs to the Sangallo brothers, notably in the fortress of Nettuno some 60 km south of Rome, but the Venetian engineer Sammicheli first made widespread use of it in defensive works. In the fortifications of his home town of Verona, he built low-lying bastions and curtain walls, with revetments of brick masonry strengthened by interior counterforts and a thick earthen rampart. In the service of the Venetian Republic he built fortifications for many of their mainland possessions and claimed that his fortress at Padua was the strongest in all Christendom. He also travelled as far afield as Crete to build defences against the Turks.

As well as being a builder of fortresses, the French engineer Vauban took part himself in at least fifty sieges. In order that troops could advance, albeit slowly, on a walled fortress, Vauban conceived the technique of excavating narrow trenches that advanced towards the walls in a series of short zig-zags, which nullified the possibility of cannon balls fired by the defenders hurtling along linear trenches, causing havoc and severe loss of life. Additional protective means included wheeled movable wooden shields, overhead timber roofing and gabions, or baskets of earth to raise to a higher level the embanked earth alongside the trench, coming from the trench itself.

Vauban's defensive works included a chain of forts to protect France's north-east border lands. He did not, by his own admission, blindly follow inflexible rules and systems for his fortress constructions, but adapted his military architecture to site conditions, and he continued to develop his ideas throughout his working life in the second half of the seventeenth century and into the early years of the eighteenth century. Notwithstanding this, his pupils and followers did contrive to compartmentalise his defensive works broadly into three systems, each, for the most part, consisting of a basic polygonal inner walled shape, the differences arising from the nature of the outworks employed, which resulted, in some cases, in quite complex star patterns.

He applied his 'first system' in the construction of some thirty out of a total of 150 fortresses in which he was involved. This system relied on well established practices, utilising short curtain walls to establish a polygonally-shaped fortress, with additional low-lying protection works (tenaille) immediately in front of each curtain wall, connected by a covered passageway (caponier) to a ravelin further out in the ditch. Generous bastions occupied each corner of the polygon. Further protection was afforded by a sloping earthworks (glacis), topped by the outworks of the fortress and running down to meet the natural countryside.

The 'second system' had added ditches and walls to increase the depth of the defence, but the most important feature was the detachment of the corner bastions from the curtain walls themselves, separating them by a ditch. This design arose from Vauban having observed that the capture of an attached bastion inevitably led to the fall of the citadel. He replaced the bastions at the corners with two-storey polygonal towers with embrasures through which cannon could be fired.

The 'third system', at Neuf Brisach on the left bank of the Rhine, some distance back from the river, comprised a synthesis of his ideas and methods, made possible by his commission to build an entirely new fortified town from scratch. The town itself, occupying flat ground and enclosed within an octagonal plan, consists of a regular grid of streets and a central square about which the important civic building are clustered. It presents an equally fortified defence from external aspect, with increased depths of the defences and added additional outworks to protect the bastions and ravelins. Bastion towers, fronted by detached bastions, occupy each corner and returning angles halfway

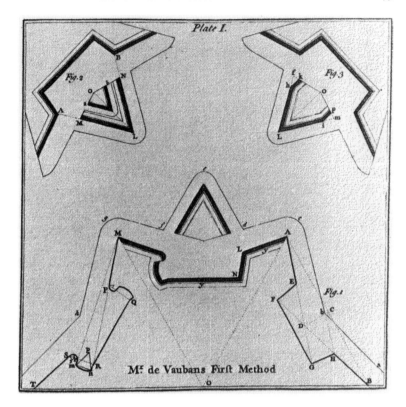

Vauban's first fortress system.

Vauban's third fortress system.

along each wall allow full flanking fire. These works at Neuf Brisach are still intact and can be seen today. The town itself is still populated.

Louis XIV's able and ambitious Minister of Finance, Jean-Baptiste Colbert, recognised fully the importance of good harbour facilities in promoting and maintaining trade. He also recognised that the major port of Dunkerque, a haven for pirates, needed to be improved to take ships up to 800 tons and at the same time needed stronger fortifications to see off possible attacks by the British from the sea and the Dutch from the land. He gave the responsibility for this enterprise to Vauban and set him to work on it in 1678. The completed fortification works proved for thirty years to be a thorn in the side of the British, who ordered them to be dismantled in 1713 as part of the Treaty of Utrecht. Louis XV later had them reconstructed. Vauban built an entirely new harbour with ships' basins and quays next to the town; as these were separated by some 2 km from the sea by a sandy shallow beach, to access them he excavated a broad channel through the sand, flanked on either side by high revetted embankments with roadways along the top. Two forts guarded the sea entrance, and about two-thirds of the way from the town he built a famous round fortress, the Risban, founded on piles driven through the shifting sand bar. Realising the need to keep the channel clear from silting up, Vauban built dams above the town to collect water by means of four canals from nearby marshes. Opening the sluices from time to time allowed any sand that had collected in the channel to be driven out to sea.

8

Towers Vertical and Towers Leaning

Towers served many purposes throughout the pre-industrial age, both secular and religious, the latter category including Christian bell towers, Islamic minarets and the stupas, temples and pagodas associated with Hinduism and Buddhism. In the middle ages the hilltop town of San Giminagno in Italy had a multitude of towers, said to have numbered as many as seventy two, some over 50 metres in height, of which some fourteen remain today; their original purpose may have been little more than as status symbols erected by its wealthy merchants, bent on outscoring each other. The town derived much of its wealth in the thirteenth and fourteenth centuries directly from its location on the Via Francigena, a busy route for traders and pilgrims travelling to and from Rome or the Vatican. There is little evidence that the towers had any military purpose, even though internecine fighting did take place in the town between Ghibelline and Guelph factions in the thirteenth century. It has also been suggested that wealthy textile manufacturers built the towers to hang their precious cloths, treated with the locally secret, and closely guarded, yellow saffron dye to protect it from the sun and dust. The longer the cloth, the greater the value; so the higher the tower, the more apparently wealthy the merchant.

The Hanseatic League spurred sea trade in northern Europe and the need for coastal towers for lighthouses and in some cases as a protection against pirates. A massive 35 metre high square brick tower built in 1369 on the island of Neuwerk at the mouth of the River Elbe, which replaced an earlier structure destroyed by fire, still stands as Hamburg's oldest building and last fortification. The tower had no light until 1644, when a 23 metre high timber structure was built on top of it to support a charcoal-fired blinker. The beacon now in use on the structure dates from 1814.

Cordouan and Eddystone Lighthouses

After joining together near Bordeaux, the waters of the Garonne and Dordogne rivers flow into the treacherous estuary of the Gironde, which is cluttered with rocks and shoals and receives the full impact of swells generated by storms in the Bay of Biscay. One of the shoals is the islet of Cordouan, which is submerged at high water. On this ledge of rock, between 1584 and 1611, the French architect/engineer Louis de

Fourteen towers now remain of the reputed seventy-two, some over 50 metres in height, in San Giminagno, Italy, in the middle ages.

Foix built the most remarkable lighthouse since the Alexandrian Pharos and which became the forerunner of the modern rock lighthouse. It replaced a smaller adjacent structure built in the fourteenth century which had fallen into disrepair. In order to overcome the reluctance of traders to brave the estuary to buy shipments of their wine, the good citizens of Bordeaux had erected some form of beacon towers on the islet from at least as early as the ninth century. In the fourteenth century, while governor of Bordeaux, Edward the Black Prince, son of King Edward III of England, had a 15 metre high structure erected on the islet, topped with a platform on which a fire could be kept burning, originally tended by a religious hermit, who may well have jumped at the job when offered it. All ships passing the light had to pay a levy of two groats.

On a solid base 40 metres in diameter and 2.5 metres high, de Foix built a three-storey structure, on top of which he placed the lantern, nearly 50 metres above sea level. The structure itself was an architectural creation generously bedecked with statuary, carved work, arched doorways and gilt. The lowest floor, 15 metres in diameter, housed apartments for four lightkeepers and included a lavishly decorated entrance hall nearly 7 metres square and 6 metres high, the second storey a royal suite probably never used by any monarch and the third floor a domed chapel beautifully adorned with mosaics, but rarely likely to have been blessed with clergy to preach or a congregation to listen.

Above the chapel roof, the upper portion consisted of two further galleries and the lantern structure. Doric pillars ornamented the exterior of the lowest storey, Ionic and Corinthian the next two and a mixture at the top, so numerous that they restricted the light itself, leading to replacement of the lantern house in 1727, during the reign of Louis XV, by an iron structure.

The portion of de Foix's structure up to, and including the chapel, still stands but denuded of its effigies of monarchs and nobles, removed by vandals during the Revolution. The portion of the structure above the chapel, now rising to nearly 60 metres above sea level, was completed in 1789 by Teulère, chief engineer to the city of Bordeaux, giving it a less romantic appearance, but one that accords much more closely with modern lighthouse design.

While over-elaborate to an almost absurd degree, nevertheless de Foix's solid structure provided the stimulus for the construction of lighthouses on sea-swept rocky outcrops. It even made it feasible to contemplate the construction of a lighthouse on the dreaded Eddystone rocks, which lay in wait for merchant ships sailing to and from Plymouth Harbour on the south-east coast of England and claimed many victims. Early representations by leading Plymouth citizens to Trinity House, a corporate body of Brethren charged with the responsibility for the management of lighthouses in England and Wales, fell on deaf ears, but the continual loss of valuable cargoes from America and the Far East and the decision by William of Orange to establish a major naval base at Plymouth eventually forced their hand. There only remained the problem of finding someone to take on the task. Who would be sufficiently daring, crazy enough even, to attempt to build a lighthouse tower on a small piece of treacherously sloping rock that only projected above water level for a few hours each day, ten miles from land and pounded by Atlantic winds and seas! His name was Winstanley. Although not crazy, he was one of history's characters. Having lost two of his own ships on the rocks, he possessed a strong motivation to attempt the task.

A man of volatile mind and mechanical bent, Winstanley filled his own house with ingenious devices, illusions and apparitions and threw it open to paying visitors. He also opened a place of entertainments in Piccadilly, London, called 'Winstanley's Waterworks', which featured many trick effects employing water spouts and jets and the mingling of fire and water. It remained a popular hit for over thirty years. His closest approach to building construction before Eddystone was in the exercise of his draughting skills to produce a set of drawings and sketches of Audley End Palace in Essex, where he started his working life as a junior assistant in the Estate office.

Winstanley decided to anchor his tower to the outcrop by means of twelve iron bars fixed into the rock. In the limited working hours available, it took the brawny workmen of Cornwall and Devon, wielding handpicks, the whole of the summer of 1696 to make the holes in the brutally hard gneiss; their gut-breaking efforts being finally rewarded in late October when they placed the tall iron stanchions in their receiving holes and poured in molten lead to secure them to the rock. The following summer saw the completion of the solid base of granite blocks cemented around the iron bars, which he enlarged at the start of the summer of 1698 to a diameter of 16 feet (5 metres) and a height of 18 feet (5½ metres). A slight setback occurred during work in 1697 when the French

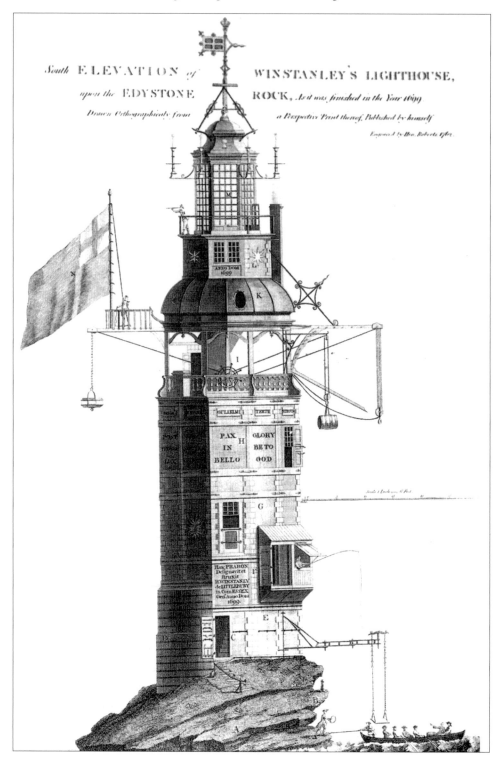

Although it was destroyed by the Great Storm of 1703, together with its builder, Winstanley had demonstrated that a lighthouse could be built on the Eddystone rock. Furthermore, he demonstrated its effectiveness as no ships foundered on the rocks in its five years of operation.

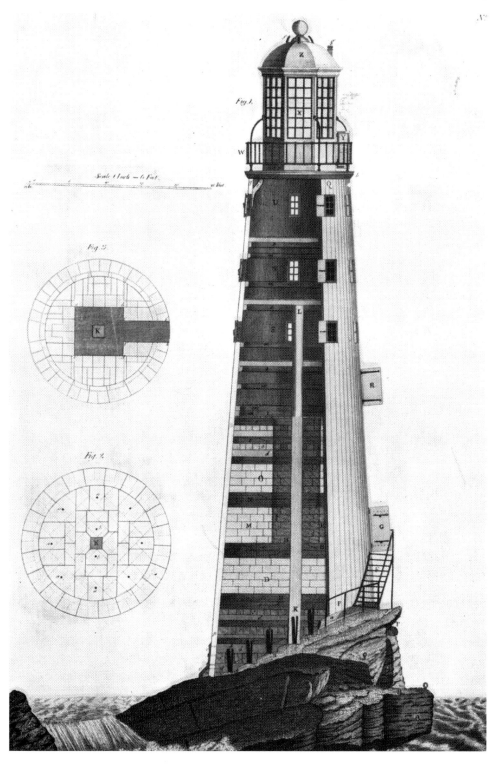

Constructed largely of timber and completed in 1708, Rudyerd's Eddystone lighthouse stood until 1755, when it succumbed to fire.

navy seized Winstanley and took him to France, but when Louis XIV heard of this he had him quickly released, despite the war between England and France.

By the end of the 1698 working season the stone portion of the tower had been completed to its intended height of 35 feet (11 metres), with an octagonal section above the solid stone base containing the living quarters of the lighthouse keepers. A timber structure above this comprised an open gallery with an umbrella roof, the lantern chamber and a weather vane 80 feet (24 metres) above sea level supported by an elaborate wrought iron bracket. On 14 November 1698, Winstanley first lit the tallow candles and people flocked from afar to see them. In the following five years not one ship foundered on the Eddystone reefs.

After the first winter, Winstanley noticed that much of the cement between the granite blocks had washed out under the relentless pounding of the waves. He immediately resolved to encase the whole of the tower in a new one, increasing the solid base to a diameter of 24 feet (7½ metres) and the height to 20 feet (6 metres) and increasing the overall height of the stonework to 60 feet (18 metres), with the weathervane now 120 feet (36 metres) above sea level. He covered with iron bands the joints that received the main force of the waves. The enlarged tower boasted a 'State Room', probably inspired by the Royal Suite in the Cordouan Tower, which Winstanley would certainly have known about, and various appurtenant features, including two projecting cranes and a projecting gallery from which signals could be made and from which the ensign could be hung.

Speculation in Plymouth at the end of each summer as to whether or not the tower would survive another winter prompted Winstanley to remark publicly his great wish to be in the tower during the greatest storm there ever was. Mother Nature clearly saw this as a challenge and decided to grant him his wish. The greatest storm ever to hit the southern half of England came on the night of 26/27 November 1703. One hundred and twenty three people were killed, 800 houses completely destroyed and many thousands severely damaged, 400 windmills blew down, at least 100 churches lost their roofs and seven spires collapsed. Hundreds of thousands of trees were uprooted. One hundred and fifty ships sank in huge seas.

Bad weather for several days before the final cataclysm had prevented Winstanley from going out to the lighthouse to effect some urgent repairs. On the morning of Friday 26 November, the weather relented briefly and Winstanley seized the opportunity to go out to the rocks despite the rough seas. Unable to get off again, he perished in his beloved tower when it simply disappeared in the final screaming fury of the storm. As a structure the tower had its faults, but it was a cruel stroke that it had to face an unprecedented storm shortly after its construction, otherwise it may well have lasted many years with suitable maintenance. Unarguably, Winstanley had conquered the Eddystone reef, and it could be done again.

In 1706 a Cornishman took up the challenge. Little is known about John Rudyerd except that he fled as a boy from a brutish father and bullying brothers and found, providentially, a kind employer who had him educated and set him on a business career that led eventually to his becoming a silk merchant. He reasoned that as ships were made of wood and survived constant battering by waves, this had to be the logical material for a lighthouse. He had the sloping rock of the Eddystone outcrop cut into a series of steps to provide a better foundation and sank thirty-six iron rods into the rock to tie the tower immovably to the reef. He then built up a solid base 9 feet high comprised of alternating layers of timbers and granite blocks, tied together with iron strips, bolts and clamps, and above this a semi-solid section 18 feet high of similar construction, but with a central shaft for stairs and an access door. Above this the construction consisted entirely of timber, bringing the final height to 71 feet. The whole of the tower below the lantern was encased in vertical timber boarding 9 inches thick, fashioned from the finest Devonshire oaks, to give a smooth tapering exterior 22.7 feet diameter at the base and 14.3 feet at the top. All joints were well caulked with oakum and the whole exterior liberally coated with pitch. Completed in 1708, the lighthouse operated satisfactorily, despite woodworm attack in the lower timber which had to be replaced, until 1755 when, in December, it succumbed to fire, the constant enemy of timber structures. An apparently retiring man, John Rudyerd appears to have disappeared completely from public view immediately after completion of the tower.

The destruction by fire of Rudyerd's tower opened the way for John Smeaton to build his great structure on the Eddystone rocks. He received his commission to build the structure in 1756, aged 32, from Robert Weston, the principal shareholder of the small group of shareholders who held, from Trinity House, a 99-year lease on the site dating from 1708 which allowed them to collect dues from shipping in return for maintaining a light on the Rock. For most of his working life up to this time Smeaton had been a maker of scientific instruments, in which field he was held in sufficiently high regard to be elected a Fellow of the Royal Society at the age of 29. However, in 1752 his interests turned to the power of wind and water, on which he carried out a number of classic experiments, a field clearly relevant to lighthouse design, although he may not have had this specific application in mind when undertaking these studies.

Up to the time of receiving his commission from Robert Weston, his civil engineering work had been limited to a few watermills and windmills and some proposals for bridge designs and land drainage schemes. He had also made a five-week trip to France, Belgium and Holland to study civil engineering works, particularly hydraulic schemes, in those countries. In view of Smeaton's lack of experience in the design and construction of major work, it redounds much to the credit and foresight of Robert Weston that he accepted the warm recommendation of the Earl of Macclesfield, President of the Royal Society, that he was the man for the job.

He decided from the outset that the tower's structure should be entirely of stone. This despite the fact that Rudyerd's largely timbered structure had successfully defied for nearly fifty years everything the winds and sea could throw at it, succumbing in the end not to these savage forces, but to fire – the scourge of timber. He dismissed the argument that timber should be adopted, because its flexibility would allow it to sway and survive the forces of wind and water, as such large movements or agitations of the structure were in themselves undesirable. He envisaged a building of such weight and solidity that instead of it giving way to the sea, the sea would have to give way to it. And, to boot, a structure not so susceptible to destruction by fire.

Having had no previous experience of the site, and indeed never having been to the West Country before taking on this commission, Smeaton wisely took on as his assistant Josias Jessop, a shipwright who had for many years been responsible for the maintenance and running repairs of the previous lighthouse. He also proved himself a fine craftsman, and skilled draughtsman and modeller. Their working relationship proved very successful once Smeaton had managed to persuade the much older man that the best solution for the structure was to construct it of stone and not timber, on the basis not so much of stability, about which Jessop still felt a flexible tower could be more stable, but on the problems of fire and woodworm, the latter having attacked Rudyerd's tower to the extent of requiring replacement of some of the woodwork.

Although Smeaton had never seen Rudyerd's lighthouse, he was able to study models and drawings of it provided by Weston, and thus assess why it withstood everything the winds and waves could throw at it. He concluded that the concentration of weight at its base contributed much to this stability, which he himself could exploit by adopting a curved, tapering profile, rather like a tree trunk, with a broad, heavy base.

Having regard to the hazardous site conditions and difficulty of access, but at the same time wishing to keep site handling to a minimum, Smeaton concluded that the optimum size of stone blocks would be 1 to 2 tons. He further concluded that the more difficult to work, but harder, Cornish granites should be used for all blocks exposed to the elements, but softer Portland stone could be used elsewhere. Wave damage was not the only potential problem with Portland stone. Below the low-water mark, the stone could suffer damage from a small crustacean boring holes into it.

Smeaton realised that the steps cut into the rock had contributed much to the stability of Rudyerd's structure and adopted the same foundation strategy, but cut the steps with much more precision than Rudyerd had achieved. Each step projected slightly above its occupant layer of blocks to give resistance to sliding. In order to achieve the required precision, all cutting had to be done by hand with no blasting by gunpowder. The circular arc of stone blocks increased in length with height up to the sixth course, which became the first complete circular level platform. The blocks were dovetailed together, a technique more common in mortise and tenon joints in carpentry, and also dovetailed into the rock in the lowest foundation courses. The joints in each course were staggered laterally relative to the joints in the layers immediately above and below. Courses were pinned either to the rock or to each other by trenails of very hard oak timber driven into matching holes to resist uplift by wave action, and additional resistance to sliding provided by inserting marble cubes or joggles into matching recesses in the top and bottom faces of successive courses.

Courses fifteen to twenty-four contained the entrance passage and stairs and above this the solid construction gave way to circular walls, one granite block thick, the blocks no longer dovetailed but held together by iron cramps. The walls supported vaulted flooring at three levels and an upper cornice and vaulted roof, which, in turn, supported the lantern. Iron ties from the iron frame of the lantern extended down into the tower walls. Outward thrusts from the vaulted floors against the walls were contained by circular chains at each floor level.

Smeaton conducted extensive tests to find a suitable mortar for the rock joints, and eventually fixed on a cement combining lime derived from a siliceous limestone from

Aberthaw in South Wales and pozzolana from Civita Vecchia in Italy, a substance well known to the Romans for its properties of setting under water. In order to protect external joints against wave and tidal action during the setting process, Smeaton covered the outer edges of the joints with quick-setting plaster of Paris.

Construction on the lighthouse commenced on 5 August 1756 and the first light from Smeaton's tower shone out on 16 October 1759. It stood for over 120 years, the last light showing in 1882, to be replaced by a light with a much greater range shining from a tower twice as high, founded on a nearby, more sheltered, reef.

Dismantling of the tower revealed only minor structural weaknesses in the original structure, such as some tendency to uplift of the upper joints under the action of waves impacting against the cornice, which had required the addition of vertical ties. Its profile and principles of construction served as icons for modern lighthouse design. A stump of Smeaton's tower can still be seen.

Galata Tower

Towers built for military purposes usually formed an integral part of a defence system of walls, ramparts and towers, perhaps surrounded by a moat. They gave defenders a commanding position from which to rain down arrow or other missiles, molten lead and boiling oil on besiegers. Towers placed on strategic bridges served similar purposes. Such towers also served as lookouts, and some were built primarily or solely for the purpose, as in the case of the Galata Tower in Istanbul, occupying a hillside site with an elevation of 35 metres above sea-level, 425 metres from the northern shore of the Golden Horn. Built by the Genoese in 1348 to protect their colony at Galata, it is ideally situated for the purpose, commanding a panoramic view over Istanbul and the Bosphorus.

The tower rises through nine floors above the high-ceilinged entrance hall, to attain a total height of 62.6 metres to the top of the conical roof. It has a base width of 16.45 metres and the dressed stone walls are 3.73 metres thick; several rows of bricks incorporated into the structure at second storey height and at 17 metres height are a typical Turkish feature that may have been added during reconstruction of the tower after a major earthquake in 1510. Despite various reconstructions, the external appearance today is thought to be much like the original Genoese construction. Most of the windows in the lower floor, apart from a window above the entrance door, are mere slits, but increase in size higher up at sixth and seventh floor levels. Large round arch windows encapsulate the eighth floor and smaller ogival arch windows the ninth floor, which now boasts a restaurant and night club and is ringed outside by an observation deck.

The tower has seen various restorations over the centuries, in addition to that in 1510: notably after a major fire in 1794 in the reign of Selim III, in 1832 in the reign of Mahmut II, who thoughtfully had the poet Pertev inscribe sixteen lines in Arabic calligraphy above the ornate entrance door praising him (Mahmut) for the restoration, and most recently between 1964 and 1967 when major changes to the interior included erecting a concrete structure incorporating a lift, and newly installed stairs replacing the original wooden steps, allowing tourists to reach the restaurant and nightclub by stairs from the penultimate floor.

The Galata Tower in Istanbul was built in 1348 by the Genoese on a prominent hillside site to guard their colony.

The many uses of the tower over the centuries have included the housing of the Janissaries, the elite of the Ottoman army, as well as, in the sixteenth century, the housing of prisoners of war destined ultimately to become galley slaves. The astronomer Takiudin used it as an observatory in the sixteenth century during the reign of Selim II, but the intention of Feyzullah effendi to make similar use of it over a hundred years later, during the reign of Mustafa II, with the help of a Jesuit priest, came to an end with his death in 1703. In 1638 Hezarfen Ahmet Çelebi is said to have become the first aviator when he took off from this tower with artificial wings and flew across the Bosphorus to the slopes of Üsküdar on the Asian side. A relief on the seventh floor commemorates this exploit. In the period before its most recent restoration, which started in 1964, it was being used as a fire-control station. It is now open to tourists, diners and night-clubbers.

Stupas and Pagodas

Many towers, in various guises, have been built for religious purposes, in some cases free standing and in other cases as an integral part of a church, mosque or temple. The Hindu and Buddhist faiths have both produced stupendous numbers of masonry stupas and tower-like temples throughout India and south-east Asia. In some a linear tapering form has been favoured, while others display a parabolic profile. Most have lavish carvings and ornamentations embellishing their exterior surfaces. A large stupa at Bodh Gaya in eastern India is the most important shrine of Buddhism. Construction on the site started in the first century BC, with the stupa reaching its final structural form at the time of Gothic construction in Europe, but it did not attain its final structural form until the seventeenth century AD. It has inspired many other Buddhist temples, its veneration stemming from its location at the place where the Lord Buddha, around 500 BC, achieved enlightenment while sitting in contemplation under a banyan or fig tree.

An impressive eleventh-century tower-temple dedicated to the worship of Shiva, in some Hindu denominations the God of destruction, one of the trinity of events in life, can be seen in Thanjavur, a silk-producing city about 300 km south of Madras. The 63 metre high tower, set on a 14 metre square base, is lavishly decorated with the heads of lions and crocodiles, as well as other imaginary animals, and with gods suitably attended by their women. A huge stone 80 tonnes in weight caps the apex of the tower, the placing of which was clearly an impressive engineering feat.

As Buddhism spread to China, Korea and Japan, these countries became adept at building their own forms of pagodas, with both brick and timber commonly used in their construction. The 54 metre high Iron Pagoda in the Chinese city of Kaifeng, about 650 km south of Beijing and built during the period of the Northern Sung Dynasty in 1044, is a twelve-storey structure of brick with iron coloured tiles (hence its name) covering its octagonal exterior. It has successfully survived many earthquakes, typhoons and floods. The Japanese mastery of carpentry, probably learnt from the Koreans and Chinese, is well attested to in their many temples and pagodas, but perhaps nowhere better than in the 10 metre square, five-storey pagoda built in 607 AD to a height of 33 metres within the Horijuji Temple complex.

Minarets

The slender and graceful minarets that are an invariable feature of Islamic mosques originated in Syria under the Umayyad dynasty, but appear to have evolved initially from the four ancient corner towers of the first century AD Greek temple and tenemos (sacred area around the temple) at Damascus. The Syrian Caliph al-Walich I chose this site on which to build the Great Mosque of Damascus, a fitting monument to the man who extended the Islamic empire to Spain in the west and deep into India in the east.

Some early minarets more resembled watchtowers architecturally than the elegant structures so familiar today. The ninth-century Malwiya minaret in Samarra, a holy city on the Tigris River 100 km north-west of Baghdad, is a 50 metre high red brick tower standing on a 33 metre square base. Helicoidal in shape, tapering towards the top, its external profile is dominated by the 2 to 3 metres wide spiral ramp running almost the full height. The minaret of Ibn Tulun in Cairo bears some similarity to the Samarra minaret for the reason that its eponymous builder, the son of a Turkish slave, was born and brought up in Samarra. Finding favour with the Caliph, he rose to become Governor of Egypt in 869 AD. The lowest part of the minaret, which is square in plan

Completed in the early thirteenth century, at the beginning of Muslim rule in India, the Qutab Minar in Delhi rises to a height of 72 metres from its 14 metre diameter base, tapering to less than 3 metres at its apex.

and 21.5 metres high, is surmounted by a circular shaped storey and this in turn by two octagonal shaped storeys. The tower tapers inwardly with height and a rising staircase that surrounds the lower square section continues in a half turn around the circular section. The present minaret dates from 1296, when the original brick structure, having suffered damage, was rebuilt in its original form using limestone.

All Syrian minarets built before the thirteenth century were square, the five-storey minaret of the large mosque in Aleppo, completed in 1094, providing an excellent example. For the most part, however, the circular form prevailed and spread from Iraq through Persia to central Asia. An imposing example of a free-standing tower and one of the finest monuments in India is the Qutab Minar in Delhi. While certainly used by the muezzins to call the faithful to prayer, it may also have been built as a victory tower to mark the beginning of Muslim rule in India and no doubt it also served as a watchtower. In 1193 the brilliant Turkish general Muhammed Ghuri finally overcame fierce Rajput resistance and captured Delhi, establishing there a Turkish Muslim dynasty which lasted 300 years until 1526, when Delhi fell to another Turkish general, Babur, who founded the Moghul Empire. Qutabud din Aibak, a former slave of Muhammed Ghuri, who became the first Muslim ruler of Delhi, commenced construction of the Qutab Minar in 1200 AD, but it had only reached one storey in height by the time of his death in 1210. He died as a result of falling from his horse while playing polo. During the Sultanate of his successor Iltutmish, who consolidated Turkish dominance in northern India, the tower rose by another three storeys, its fluted tapering profile faced with red buff sandstone richly sculpted with plant-like decorations and Arabic letters depicting passages from the Koran. Three more marble-faced storeys followed but two storeys fell, leaving a five-storey structure with four projecting balconies supported by elaborately decorated brackets. The 72 metre high tower has a width of more than 14 metres at its base, which diminishes to less than 3 metres at its apex.

Bell Towers

The Christian bell tower had its genesis in Ravenna, which became the Byzantine seat of government in Italy after its conquest by Justinian's general Belisarius in AD 540. The famous Christian basilica of Sant' Apollinare in Classe dates from the early Byzantine period, but its associated campanile dates from the tenth century, by which time Ravenna and its port Classe had become territories of the Papal States. Although this Ravenna tower is circular in plan, square or rectilinear towers became much more the norm. Giotto adopted the square form, with 15 metre sides, for his campanile of Santa Maria del Fiore in Florence. Vasari quotes Ghiberti as saying that Giotto prepared models of the structure itself, and for some of the scenes in marble that adorn the faces of the tower. High towers are particularly sensitive to foundation movements and Giotto may well have realised this because, according to Vasari, he excavated to a depth of 12 metres and after pumping out the water placed, firstly, some 7 metres of good ballast and above this 5 metres of masonry to support the 88 metre high tower. Work commenced in 1334 under the direct supervision of Giotto himself, his having been given Florentine citizenship and granted a fee of a hundred gold florins a year, but he died in 1336 before completion of the tower and, according to Vasari, construction was brought to a conclusion by Taddeo Gaddi, who later built the Ponte Vecchio in Florence.

The sensitivity of towers to foundation movement arises because the tilting of a tower produces high stress concentrations in the ground beneath the foundations and also within the masonry of the structure and the foundations. The concentrated localised stresses imposed on the soil can, over a period of time, lead to further differential settlements, causing exaggerated tilting and even higher localised stresses in the soil and the masonry. Masonry, too, has time-dependent characteristics and can deteriorate under high stress concentrations, leading to cracking or crumbling.

Many bell towers have collapsed, but none more catastrophically than the original San Marco tower in Venice. In the spring of 1902 some fissures and a large crack appeared in the masonry, which increased visibly between 11 and 13 July; then on 14 July the crack opened dramatically and the tower collapsed into a heap of rubble. Despite the notoriously bad soil conditions in Venice, the tower had developed only a slight tilt, insufficient to cause any instability, suggesting that in this particular case deterioration in the bricks and mortar making up the structure led to its collapse. The original tower rested on a foundation consisting of five courses of stone with a stepped profile supported by a plinth of roughly squared stones, which, in turn, rested on a wooden slab and timber piles driven down to a cemented silt layer.

During its lifetime of more than one thousand years, the tower witnessed many great events in Venice, which, from 1095, cashed in on the Crusades – building and outfitting ships, equipping knights and charging outrageous prices for voyages to the holy places. For 500 years it was the most powerful maritime state not just in the Mediterranean, but in the whole of Europe. It developed its own industries, notably in textiles, lace, glass and shipbuilding. When victorious armadas returned home, the bells of the campanile rang out over the whole lagoon and climbing the tower invariably appeared high on the itinerary for visiting dignitaries. Not so felicitous was the medieval custom of dangling violators of Church laws from the belfry in wooden cages.

After the collapse of the old tower the Municipality of Venice decided that its replacement '…should rise in the same place and be of the same shape and colour, even though the structure and the building technique may differ to be in compliance with modern science which may guarantee a longer life…' This has been abbreviated in modern Venetian lore to *comm'era, dov'era* ('as it was, where it was'). The new structure was completed in 1912 to the same height of 99 metres, but of lighter construction and with an enlarged foundation supported by deeper piles.

A plan view of Pavia, a Lombardian town in Italy on the Ticino River, drawn by Ludovico Corte in 1617 showed about ninety towers within the town limits. A sixteenth-century fresco still visible in the Church of San Teodoro shows about forty towers. As fifty towers are unlikely to have been built in the intervening period, it seems that the fresco painter may have exercised an artist's license by leaving a few out for clarity. In fact, a great many of the towers dated from the twelfth century, probably constructed as family symbols proclaiming family wealth, as in the case of the San Giminagno towers, and they may also have served as watchtowers. Many met their demise as a result of feudal rivalries, while others collapsed on their own account, including the belltower

Over a thousand years old, San Marco tower in Venice collapsed catastrophically on 14 July 1902, apparently from deterioration of the bricks and mortar making up its structure. It was rebuilt, identical in appearance, in 1912.

of the church of San Niccolo in 1581 and the so-called Torrione in 1712. The soil below Pavia is Ticino sand, and some of the collapses probably occurred because sand is a weak material when subjected to concentrated loads at shallow depth and also is greatly weakened by any rise in the ground water level. Perhaps the most slender and elegant tower still standing is the Chiesa del Carmine, with a height of 66 metres, side widths of about 8 metres and founded at a depth of 5.5 metres below ground level.

On 17 March 1989, the 60 metre high Civic Tower in Pavia collapsed suddenly, killing four people and severely damaging nearby buildings. It had previously shown no sign of distress or instability. Construction works on the lower orders of the tower are thought to have been started between 1060 and 1100, with higher orders added in the twelfth and thirteenth centuries and topped out with the belfry in the sixteenth century. The 2.8 metre thick walls of the 12.3 metre square structure consisted of brick facings averaging about 150 mm in thickness and in-filled with stones, pieces of brick and mortar. Extensive investigations showed the underlying soils to be mainly strong sandy deposits, ruling out the possibility of foundation movements causing the collapse. The most likely cause seems to have been a gradual deterioration of the structural materials of the tower over the centuries under its own dead load. Cyclonic winds seven months before may have accelerated the collapse, but did not cause it.

Although differing greatly both in structure and ground conditions, the collapse of the Civic Tower in Pavia served as a reminder, if any was needed, that, despite many attempts, no solution had been found and agreed upon to stabilise the most famous tower of them all at Pisa. Its fame deriving mainly from its disconcerting lean, it nevertheless has architectural merit consisting, unusually, not of a square but a cylindrical load-bearing structure of eight orders made up of the base, six levels decorated with columns and vaults, and the bell chamber.

Pisa Tower

Construction of the Pisa tower commenced in 1173, once thought to have been under the direction of Bonanno Pisano, more noted for his bronze castings than his architecture, and whose sarcophagus was found at the foot of the tower in 1820. It now seems more likely to have been under the direction of Diotisalvi, an architect with previous experience in bell tower and cathedral construction in Pisa. By 1178 the foundations and the first three orders, and part of the fourth order, to a total height of 29 metres, had been completed. Work then stopped for almost a century for reasons unknown, but fortuitous in the light of modern calculations indicating that bearing capacity failure would have occurred in the underlying soil if the tower had been taken to full height without pause. Stoppage of the work allowed dissipation of excess water pressures (pore pressures) in the ground below the foundations, generated by the weight of the structure. This process can continue over many years, leading to a gradual strengthening of the soil. The second phase of the work from 1272 to 1278 saw the tower completed, except for the bell chamber, which was added between1360 and 1370. This second delay may have been a cautious reaction to observed settlement and tilting of the structure. After construction recommenced in 1272, the tilt

The slightly curved shape of the Pisa tower reflects the fact that it was built in stages between 1173 and 1370 and attempts were made to correct the verticality at each stage as it settled to the south. It has continued to settle throughout its existence until the recent successful measures taken to stabilise it, leaving it with a lean of about 5½°.

towards the south became pronounced and increased rapidly between the erection of the sixth and seventh orders. Correction for the tilt by careful hand cutting of the stone blocks began early on in construction, with a very significant correction for the bell chamber, giving the tower a distinct curvature in profile.

Pisan builders may have been quite accustomed to settlements of structures founded on local soils. The cathedral, which is adjacent to the tower and was built before it, financed by the spoils of six large ships full of treasures taken in 1063 by the Pisan fleet from the Saracens in Palermo, shows clear signs of settlements that appear to have been corrected during construction. Another reason which has been advanced to account for the second long pause before completion of the tower was the defeat of the Pisan fleet in 1292 by the Genoese, which effectively put paid to Pisa as a major power.

The masonry walls of the load-bearing Pisa Tower cylinder consist of inner and outer concentric facings about 400 mm thick of squared, very hard, San Guiliano marble blocks with the space between infilled with rock fragments and stones cemented with San Guiliano mortar. The wall thickness changes abruptly from 4.1 metres in the first order to 2.6 metres in the second to sixth orders. Soft limestone, bricks and pumices for lightness were used in the seventh and eighth orders. The eight-storey tower reaches a height above ground level of 53.3 metres, and requires 296 steps to reach the top. The annular foundation supporting the tower at an average depth of about 3.5 metres below ground level has an external diameter of 19.6 metres and an internal diameter of 4.5 metres. With an estimated total weight of 14,500 tonnes, the tower exerts an average pressure on the soil of about 500 kilopascals. The foundations slope towards the south by about 5½°, and the seventh floor overhangs the base by about 4.5 metres.

One of the major concerns in recent times has been the structural integrity of the tower, particularly on the south side at the contact between the base segment and the first loggia. The southward tilt of the tower has produced concentrated loading on this face leading to the external marble facing of the second order sustaining high stresses, which have migrated to the much weaker infilling, because of the abrupt reduction in wall thickness from the base section (first order) to the second order. Further weakening in this section is contributed to by a large opening for the staircase. Because of the immediacy of this problem, strengthening of this section by temporary metal bands was one of the first operations put in hand in the recent comprehensive and carefully researched work carried out to stabilise the tower.

Soil conditions to a depth of some 10 metres below the tower consist of estuarine deposits generally of interbedded sandy silts and clayey silts, but with a 2 metre thick layer of fine, medium dense sand at its base. Field tests and sample descriptions have indicated the soils to the south side of the tower are slightly more clayey and silty than to the north, and the sand layer is a little thinner to the south, and that these conditions are most likely to have been responsible for the tilting of the tower to the south. A clay of primarily marine origin exists below the base of the thin sand layer to a depth below ground level of about 40 metres and below this a dense sand exists to considerable depth. Another thin sand layer occurs within the marine clay, while its upper portion consists of a soft sensitive clay known locally as Pancone clay. Investigations have shown the contact surfaces between soil layers to be essentially horizontal across the site except for dishing of the Pancone clay beneath the tower, which borings have indicated to amount to an average settlement between 2.5 and 3 metres. The water table fluctuates between 1 metre and 2 metres below ground level, depending on rainfall intensity and distribution in the area, but water pressures below this are less than static, because pumping from the sand below 40 metres depth has created downward seepage.

It can be safely assumed that over the long history of the tower, a number of unknown factors have influenced its behaviour and contributed to the settlements and shear movements in the underlying ground, causing the tower to tilt southwards to a maximum inclination of some 5½° before commencement of the recent stabilisation works. Some possible factors contributing to the tilt, such as ground water pumping in the region, are known, although not necessarily their precise contribution. In 1838 an annular excavation (Catino) made around the tower to expose the column plinths and foundation steps, and to create a walkway, caused an inrush of water on the south side where the excavation dipped below the ground water table. The tilt increased by about 0.5° as a result of this action. Further disturbance, causing a small increase of 31 arc seconds in tilt, resulted when the Catino was waterproofed, and the foundation grouted, between 1933 and 1935. Pumping from the lower sands in the late 1960s and early 1970s caused regional subsidence and induced a small tilting of the tower towards the south-west amounting to 41 arc seconds.

During the first stage of construction between 1173 and 1178 the tower showed a small tilt to the north, which increased slightly during the subsequent one hundred years of inactivity to a maximum of 0.2°. After recommencement of construction in 1272 the tower began tilting towards the south, a movement which was apparently accelerating when construction again ceased before the addition of the bell chamber; the southerly tilt now having reached about 0.6°, which increased to about 1.6° during the ninety years elapsing before the addition of the bell chamber, which triggered a dramatic increase in the inclination. Plumb line measurement made in 1817 by British architects Cressy and Taylor showed an inclination of about 4.9°, which, by the early 1990s, had increased to about 5½°, largely as a result of the Catino excavation, leaving the seventh cornice overhanging the first cornice by more than 4 metres. Eliminating perturbations from ground water lowering indicated a rate of tilt increasing from about 3 arc seconds per annum in 1940 to 6 arc seconds in 1990, leading to the realisation that without some form of stabilisation the tower would eventually topple.

Many commissions have been set up in the past to consider the stability of the tower, apparently to little effect until the sixteenth commission this century established by the Prime Minister of Italy in 1990, charged with developing and implementing measures for stabilising the tower. Its membership embraced a number of disciplines, including history of architecture, preservation and restoration of historic buildings, medieval art and archaeology, structural engineering and geotechnical engineering. Its work was hampered to some degree by the fact that the Italian Parliament did not ratify the Commission until 1995, so that prior to this its mandate had to be renewed every two months.

Early discussions of the Commission resolved that stabilisation of the tower required measures to combat localised overstressing, both in the masonry of the tower and in the soil below the foundations, which could be largely achieved by a small reduction in the tilt amounting to no more than one-half to three-quarters of a degree. Restraints were placed on the Commission in the methods it could consider, with both piling and grouting rejected out of respect for the integrity of the tower and its cultural history. Various possibilities under consideration early, respecting the need to have the least physical effect on the tower itself, included:

electro-osmosis
compressing the soil on the north side of the tower
vacuum pumping the Pancone clay north of the tower
underexcavation (removal) of soil north of the foundation and below it

All these methods presented the possibility of inducing differential settlement in the soil below the tower foundation sufficient to counter further tilting and even to reduce slightly the present tilt: the first by transferring, laterally, moisture in the soil; the second by consolidating (squeezing out some of the moisture) in the soil below the north side of the tower through loading a surface plate by means of vertical ground anchors; the third by vacuum extraction of moisture from the soil; and the fourth by the physical removal of soil below the north side of foundation.

After thorough discussion and exploration, and investigation by both physical and numerical models and, in addition, full scale tests of electro-osmosis and underexcavation, the latter of these two was the option finally adopted. The Commission subsequently discovered that the method had been used successfully as early as 1832 to stabilise the tower of St Chad in Wybunbury, in England.

The Commission decided on a two-stage approach, carrying out temporary and fully reversible measures to strengthen the structure and ensure its immediate stability, before implementing the permanent solution. The first measure, carried out in 1992, consisted of encircling the structure at four elevations on the second storey and the first cornice with prestressed steel strands to combat any tendency for the outer marble facing to buckle outwards. The second measure, carried out over a period of seven months in 1993, consisted, in a delicate operation, of placing 600 tons of lead weights to load the base of the tower on the north side. Confidence to implement this measure came in part from the discovery that the previously widely held view that the continuing tilt was caused by foundations on the north side settling less than on the south side was incorrect, that in fact the north side had been rising. This also led the geotechnical engineers on the Commission to identify a seasonally fluctuating water table as the cause of the continuing movement of the tower. The tower moved northwards by 48 seconds of arc, equal to about 14 mm at the top of the tower, and, most importantly, reduced the overturning moment by about 10 per cent as a result of placing the 600 ton weight.

At this time a potentially catastrophic blip occurred when some members expressed concern about the future of the Commission and visualised its possible disbandment, leaving the lead weights in place, and they proposed replacing the unsightly weights by ten ground anchors acting through a concrete ring at foundation level. The Commission, by a majority vote, agreed. Excavation of the waterlogged soil to receive the concrete ring presented a difficult problem, and the introduction of ground freezing to expedite this

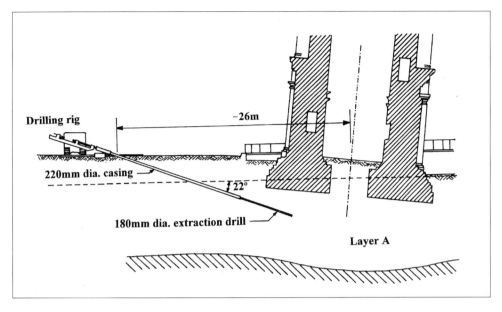

The Pisa tower has been successfully stabilised by the careful extraction of soil under its northern side by means of an inclined hollow-stemmed continuous flight auger.

resulted in additional southward movements of the tower and had to be abandoned. An additional 300 tons of lead were added to the north side to halt these movements.

Preliminary tests extracting soil from below a trial foundation in 1995 using an inclined hollow-stemmed continuous flight auger inside a contra-rotating casing 180 mm in diameter proved successful, and preliminary implementation on the north side of the tower itself commenced in 1999. As a temporary safeguard 100 metre long cables extending to the north were attached to the tower, the free ends with adjustable weights attached passing over pulleys attached to massive A-frames. With the successful conclusion of the closely monitored preliminary removal of soil below the north side of the tower foundations, resulting in a stable northwards rotation of the tower by about 130 arc seconds, full and final extraction commenced in February 2000, consisting of forty-one extraction tubes at 0.5 metre spacing. This continued until June 2001, with progressive removal from the tower of the lead weights. The total northwards rotation due to soil extraction reached a stable value of about 1,730 arc seconds or almost half a degree, equating to about 160 mm settlement on the north side of the tower.

Some additional works were carried out to further ensure the stability of the structure, including strengthening of the tower walls by grouting of voids in the rubble core and replacing the original steel tendons with less obtrusive prestressed wires set in resin. A concrete ring encircling the foundations, placed in 1838, was securely bonded to the foundations to give added bearing area and a system of controlled drainage wells was installed in the ground to eliminate continuous fluctuations in ground water levels.

Since completion of these works and the official re-opening of the tower to the public on 15 December 2001, the continuing monitoring has detected nothing more than a slight rotation to the north of a few arc seconds.

9

Soaring Vaults and Flying Buttresses

Romanesque

Although Charlemagne accomplished little in the way of building, he strengthened the hold of Christianity in Europe and paved the way for the upsurge of building led by the Church with its construction of monasteries, churches and cathedrals, firstly in the Romanesque (Norman) style and later in Gothic.

Romanesque construction during the eleventh century comprised to some extent a relearning process, as many of the skills of ancient Rome had been forgotten. In plan, the churches took up the form of the Roman basilicas – rectangular with central naves separated from flanking aisles by a row of cylindrical columns down each side supporting the clerestory. Early structures had flat timbered roofs, but these gave way in some later churches to masonry barrel vaults over the nave and groin-vaulted side aisles. Rounded arches were used throughout over window and door openings and for supporting the clerestory. The buildings usually featured a half-rounded apse at one end and an entrance structure at the other. Transepts also became a feature of Romanesque churches and, with this addition, the plan layouts of the later major churches corresponded closely to, and became the accepted pattern for, the great Gothic cathedrals. Heavy walls, in some cases with buttressing, took the lateral thrusts from nave and aisle vaults, thus restricting the size of openings and so limiting the amount of light entering the interior.

Much of the credit for setting in motion the momentum in church building from the second half of the tenth century onwards belongs to the Cluniac monks, a Benedictine order first established at Cluny, a small village about 100 km north of Lyons, when William the Pious, Duke of Aquitaine, on his death in 915 left his hunting lodge to the Church. Under the third head of the order, Abbot Hugh, who ruled for sixty years, the spiritual influence of Cluny spread as far as northern Spain, Italy, the Alps and England, and led to the construction of more than a thousand monasteries. Abbot Hugh rebuilt and enlarged the existing monastery on the Cluny site between 1084 and 1121 to create Cluny III, an enormous building unequalled in size until the completion of St Peter's in Rome nearly five hundred years later. His reason for building such a grandiose structure sprang from the need to house up to 2,000 visitors from all over Europe, in addition to housing the Order's 400 monks. Perhaps, too, he sought to rival the magnificence of the Hagia Sophia: in sheer size, his 180 metre long and 30.5 metre high structure certainly achieved this, but it lacked the magnificent crowning dome of its rival in Constantinople.

Both architects for the building, Gunzo and Hezelun, had some pretensions to mathematical skills, and the various dimensions within the main structure and its two sets of transepts, radiating chapels and numerous towers obeyed mathematical relationships based on moduli in metres of 1.52, 2.6, 7.6 and 9.4. Double aisles flanked the nave, which was roofed with a slightly pointed barrel vault springing from thick clerestory walls. This support proved to be inadequate and flying buttresses had to be provided to prevent the walls toppling. Regrettably, this great church with its soaring space, accommodating perfectly the famous Cluniac antiphonal chants, suffered wanton destruction during Napoleonic times.

Durham Cathedral in northern England, constructed between 1093 and 1130, had the advantage of coming towards the end of the great period of Romanesque church building, allowing its designers to draw on more than a century of experience. Its fine proportions bespeak the willingness of its designers to learn from previous structures. But they also showed themselves to be highly innovative in creating the first cathedral in northern Europe to be groin-vaulted throughout in stone. Furthermore, the vaults were strongly ribbed, with the panels between the ribs consisting of heavy plaster-faced rubble, which cracked badly early in its life and had to be replaced in 1235. Slightly pointed arches span transversely across alternate bays of the nave, with their action accentuated by the rib lines being carried down the faces of the supporting piers.

Although Gothic cathedral building, when fully developed, differed quite markedly from the earlier Romanesque, the changes had more to do with space and light than with the basic plans and layout of the structures, or with the individual structural elements. Gothic cathedrals retained the basic layout of the major Romanesque churches of a long nave flanked by aisles, with transepts, apsidal ends and ambulatory, and towers. Pointed arches had already been introduced into Europe in the late Romanesque period, perhaps by returning architect engineers who had accompanied the First Crusade in 1099. A notable Romanesque example was Autun Cathedral in Burgundy, completed around 1130, with a pointed barrel-vaulted nave, groin-vaulted side aisles and pointed arches between nave and aisles supporting the clerestory walls. Rib and cross vaulting or groined vaulting was well known from Roman times and buttresses predated Gothic building. Flying buttresses appear to have been essentially the invention of Gothic builders, but the same principle was embodied in the three-level Abbey Church of St Etienne at Nevers in France, built between 1083 and 1097. The second level consisted of galleries over the side aisles, roofed with half-arches springing from the tops of the first level outer walls and resting against, and thus giving lateral support to, the third level clerestory walls flanking the nave.

Romanesque churches often formed part of a monastic establishment and in consequence usually presented an austere façade, set apart from the outside world. Gothic cathedrals on the other hand reflected the growing wealth and populations of the medieval towns. They belonged to the people. Responsibility for their construction and upkeep lay with the canons, a group or chapter of individuals deriving from the earlier body of priests which, in the early medieval period, had assisted the Bishop to administer the diocese. Gradually they assumed a more communal and secular way of life, with rights of ownership and disposals of their chattels by will. They often had the power to thwart the wishes of the bishop and sometimes received quite large incomes from the diocese, particularly in towns grown wealthy on agriculture and trade. Their responsibilities included raising and administering the funds for the cathedral building and maintenance and appointing an overseer to take overall responsibility for the work. Although for the most part capitalising on their contacts with the commercial world to raise funds, they were not above squeezing money from the bishop or taking relics on tour to collect money. The very success of some of these tours drew the attentions of the wrong sorts of people and the canons sometimes found themselves set upon and robbed.

Gothic

Credit for initiating the spirit of Gothic building belongs to Abbot Suger, who became Abbot of the Royal Abbey of St Denis near Paris in 1121. No doubt he learnt much about buildings on the many missions he made to Rome to strengthen the rather fragile ties then existing between the King of France and the Papacy. He would also have had contacts with knights returning from the Crusades and learnt much from them about building methods in eastern Mediterranean countries. His rebuilding of the choir of St Denis inspired other cathedral builders, in part because of the merits of the building techniques that allowed him to introduce large windows to allow light to stream into what had previously been a dark and gloomy Carolingian basilica, but also in part because the man himself was an important and influential figure in France, having acted as Regent of France during the absence of Louis VII on a Crusade. His humble origins, which he never concealed, served to give him both an insight into the minds of ordinary people and a feel for organising space in such a way that the church attracted people, making them feel exalted and at one with God. Banishment of the Romanesque gloom went some way towards this, as did the enlargement of the choir to embrace a double ambulatory, thus enabling the large feast-day crowds of the faithful to circulate comfortably around the altars and the great cross, and to come forward freely to venerate and to kiss the holy relics of the Nail and the Crown of Thorns. For the convenience of the priests wishing to celebrate Mass, he provided radiating chapels around the choir, which soon became a common feature in cathedral construction.

Suger envisaged a structure that made a literal truth of the words intoned in the Mass for the consecration of a church: 'This is a place of awe. Here is the court of God and the Gate of Heaven.' And only the finest building conceivable by man would do for the abbey of the patron saint of Paris, and within which lay the bodies of the French kings. Every radiating bay of the new choir, seven in all, each containing a chapel, was roofed by a ribbed vault with pointed arches, achieving a gracefulness and lightness undreamed of in Romanesque construction. The delicate framework of columns, vaulted ribs and slender arches lent a continuity to the ambulatory space rather than dividing it up, and

The choir of St Denis basilica.

the large outward thrusts from the vaults found their way to the foundations through heavy buttresses on the outside of the building.

Suger began the work on the new choir in 1140, having already made repairs to the nave and started work on a new west façade. Funds for the building work came largely from the many pilgrims who flocked to St Denis, particularly at the time of an annual religious festival centred around the relics of St Denis, and which also became the occasion for holding a huge fair. He clearly employed the services of an outstanding master mason, whose name is not known. Clerics from many parts of Europe attended the consecration of the new choir of St Denis in 1144 and returned to their own sees imbued with the new spirit of cathedral construction and with the determination not just to copy it, but to improve upon it. The next 150 years saw an astounding period of magnificent cathedral construction, surpassing anything that had gone before and which has not been matched since.

Three main structural features characterised Gothic cathedral construction: The pointed arch; ribbed cross-vaults; columns, piers and buttresses.

The pointed arch was invariably used in wall openings, and to span between columns separating the central nave from the side aisles. Originally used because of its structural advantages, it eventually came to assume ecclesiastical overtones. The structural advantages of the pointed arch included a flexible rise to span ratio, while the point of the arch could tolerate, and indeed for proper functioning required, a concentrated load bearing on it. This was important in a framework of structural members carrying wall and roof loads through the structure into the foundations. The use of a high rise to span ratio reduced lateral thrusts and gave an aesthetic appearance that appealed particularly to the French, who became almost obsessed with achieving great height and verticality in their cathedrals. By forming the pointed arch from two circular arcs with different centres on the same horizontal line, the advantage with the semi-circular arch of having standard voussoir shapes was retained, with the exception of the keystone and possibly the springing blocks.

Early Gothic constructions featured simple cross-vaults with stone ribbing along the diagonal intersection lines, and also spanning transversely across the nave between adjacent piers, thus following and emphasising the pointed profile of the longitudinal elements of the vaulting. A boss covered the intersection of the diagonal ribs. During construction, the ribs, themselves erected on light timbering, then served as a support system for the placement of the stone webs or severies of the vaults, for which only light movable timbering would then have been needed. By covering the vault intersections, the ribs made accurate fitting of the web blocks along the intersection unnecessary and thus simplified the cutting of the web stones; but they also emphasised the way in which the structural forces travelled through the intersections into the supporting piers from which the ribs sprang. In fact, in some instances, at least, their structural role may have been minimal, examples existing of ribs having fallen with the webs still standing. It is more common, however, to find ruins with the ribs standing skeletally and the webs long since fallen.

In later Gothic construction, an increasingly elaborate rib system became a common feature of the vaulting until, eventually, the webbing of the vault became almost secondary

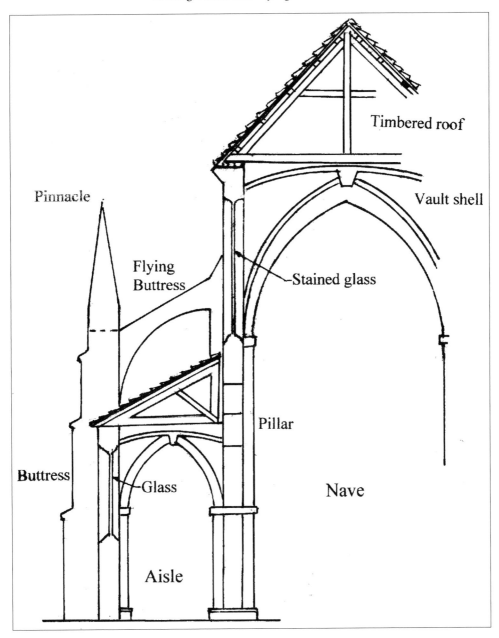

Features of Gothic construction.

to the continuous criss-cross network of ribbing. Another development of the rib system resulted in fan vaulting, such as that seen in King's College Chapel in Cambridge, England, with a fan of ribbing emanating from each pier or buttress support.

The side walls continued above the level of the vault springings to support the peaked timber framed roof, usually covered with lead sheeting. The roof served to protect the vaulting both during and after construction, as well as providing a means, by attaching pulleys to the timbering, for hauling up the blocks of masonry for constructing the vaults.

Most of the forces engendered in Gothic cathedrals by the dead weight of masonry and wind loading found their way down to the foundations through a system of columns, piers and buttresses, instead of through the walls, making them, in effect, framed structures. This, above all else, distinguished them from earlier masonry buildings. As non-load-carrying elements, the walls between the piers or buttresses could be made much thinner and the Gothic builders took advantage of this to provide large window openings, allowing light to flood into the interiors of the buildings. It made possible the beautiful stained glass windows, which, by depicting the Bible stories in magnificent colour for a largely illiterate populace, contributed much to the ecclesiastical impact of these structures. Perhaps the most dazzling display of stained glass, most of it still original, is that which contributed to making Chartres Cathedral in many eyes the most glorious and unified expression of the Gothic concept. In fact it incorporates a variety of styles. Subdued jewelled light is admitted into the interior through 124 great windows, three great and thirty-five lesser rose windows and twelve smaller ones. As befits a cathedral of Notre Dame, much of the space taken up by the stained glass exalts the life of Mary; with small details, such as the occupations of shoemakers, basket weavers, leather tanners, wheelwrights, blacksmiths and carpenters, reflecting the contributions made to the creation of the cathedral by trade-guilds or corporations. Nobles contributing to the work had their own coats of arms depicted.

Although in Gothic construction the vertical loads from the roof, the nave vaulting and clerestory walls reach the ground through the internal columns separating the nave from the aisles, these columns are nevertheless often surprisingly slender, as they are not called upon to resist any substantial lateral forces. Vertical loading from the side aisles is shared between the columns and external buttresses, the substantial nature of which reflects their primary role of absorbing the lateral forces from the roof, the nave vaulting and to a lesser extent from the aisle vaulting. In essence, these buttresses are simply short lengths of wall turned around through right angles to better absorb the lateral forces, thus allowing the walls between to be little more than weather excluding screens.

The spectacular flying buttresses, particularly noticeable in the great French cathedrals, carry the lateral thrusts from the roof and timber vaulting, together with lateral wind forces on the upper part of the structure, across the tops of the aisles to the piers or buttresses. A number of French cathedrals, such as Rheims, Bourges and Beauvais, sport two tiers of flying buttresses, the upper tier taking the lateral wind forces from the roof and the lower tier coping with the lateral thrusts from the nave vaulting. Some cathedrals, such as Amiens, Notre Dame in Paris, and Cologne, feature double side aisles requiring the flying buttresses to carry the thrusts from the nave vaulting and roof over great distances to the outer piers and buttresses. As medieval builders tended to work to simple numerical rules, doubling the length of a flying buttress meant doubling all other linear dimensions and thus increasing its volume and weight eightfold. The builders at Amiens, Cologne and several other cathedrals overcame this problem by inserting intermediate columns or buttresses between the nave walls and outer piers, with one set of flying buttresses spanning from the nave columns to the intermediate columns and a second system spanning from the intermediate columns to the outer piers. At Beauvais, Heyman estimates that the lower flying buttresses that convey the nave vaulting thrusts across the 4.5 metre span to the intermediate piers weigh about 5 tonnes each, which would have increased to 47 tonnes if they had been required to span the full 9.5 metres to the outer piers. This increase in length and weight would greatly have increased the problems of design and construction of the flying buttresses, but this did not deter the builders of the Cathedral of Notre Dame in Paris, where the spectacular flying buttresses span the double aisles without intermediate support. The single flight flying buttresses supporting the nave date from the thirteenth century and replaced the original double flight flying buttresses. The choir flying buttresses date from the fourteenth century.

An interesting and subtle feature of flying buttresses, learnt no doubt from observed disruptions or failures, was the building of extensions or pinnacles on top of the buttress piers, projecting above the levels at which the flying buttresses met them. Originally a structural device, these pinnacles developed into a decorative architectural feature adding significantly to the appearance of the structure. The pinnacles actually served to strengthen the upper parts of the buttresses against the lateral thrusts from the flying buttresses, in particular against shearing through the masonry or sliding of the masonry blocks along the horizontal joints between them, possibly to the extent of pushing the tops right off the buttresses. The extensions or pinnacles added vertical pressure at the tops of the buttresses, and thus enhanced the shearing or frictional resistance. As the mortar used in medieval times set very slowly, gradually increasing in strength over a number of years, the pinnacles were most effective and most needed immediately after construction; if the buttresses survived this initial period they could be expected to survive indefinitely in the absence of foundation movement or some other untoward happening. The pinnacles also reduced the unavoidable inclined tensile stresses near the tops of the buttresses that could lead to cracking and breaking away of the masonry.

Each pinnacle also made some contribution to keeping the line of thrust throughout the full buttress height within the middle third of its width, which avoided the possibility of tensile stresses developing on horizontal planes within the masonry. Violation of the middle third rule, however, does not necessarily in itself lead to failure of a buttress, except possibly at foundation level. As a thrust moves away from the centre, the compressive stresses increase in the direction of the eccentricity and decrease away from the eccentricity. When the load reaches the third-point, the stress at the opposite edge drops to zero and further eccentricity of loading causes this stress to become tensile. As masonry joints and contact areas between foundation soil or rock and masonry cannot resist tension, or at best can only resist very low tensile stress, an area of zero stress grows inwards from the far edge as the eccentricity increases and this causes a rapid increase in the compressive stress in the direction of the eccentricity. At foundation level, this

Flying buttresses, Notre Dame de Paris.

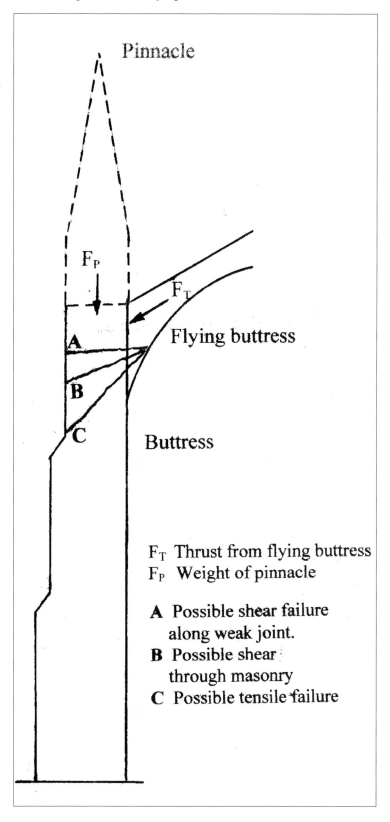

Pinnacle

F_P

F_T

Flying buttress

A

B

C

Buttress

F_T Thrust from flying buttress
F_P Weight of pinnacle

A Possible shear failure
   along weak joint.
B Possible shear
   through masonry
C Possible tensile failure

The weight of the pinnacle strengthens the junction of a flying buttress and buttress pier against the possible failure mechanisms shown.

high concentration of stress at the outer edge of the buttress may severely deform or fail the underlying soil, causing the supported buttress wall or structure to tilt or possibly collapse.

Within the masonry of a buttress, pier or flying buttress, violation of the middle third rule might lead to some minor cracking, but this is quite normal in masonry structures. If the line of thrust falls outside the masonry, then its inability to carry tension will cause the member carrying the load to collapse. A very high (theoretically infinite) compressive stress would develop where the line of thrust touched the edge of the masonry. Making the assumption that the maximum allowable compressive stress should not exceed one-tenth of the crushing stress of the masonry, Heyman suggests that the line of thrust may approach the edge to within 5 per cent of the width of the masonry before threatening the safety of a load-carrying member.

Based on this philosophy and the principles of limit design, Heyman also proposes two theorems:

> If, on striking the centring [i.e. supporting frame during construction] for a flying buttress, that buttress stands for 5 minutes, then it will stand for 500 years.

> If the foundations of a stone structure are liable to small movements, such movements will never, of themselves, promote collapse of the structure.

He goes on to qualify the first of these to apply to static loads only and not, therefore, to the upper flying buttresses of a Gothic cathedral that carry the thrusts from the wind loading on the roof. The 500 years is the assumed decay period for the stone work. The survival of so many splendid Gothic cathedrals today supports the veracity of this theorem.

The second theorem assumes that thrust lines, which are simply graphical representations of stresses in equilibrium, can adjust to stay within the masonry. It is certain that many of the Gothic cathedrals now standing have suffered substantial foundation movements, in some cases causing visible distress in the fabric, but not collapse, thus supporting the theorem. A case in point is York Minster in England, completed in 1472. An examination in 1966 of cracked and distorted masonry around the central tower revealed that differential settlements up to 225 mm had occurred between the tower and the nave, arising from the compression of the firm to stiff clays underlying the foundations. Remedial measures consisted of encasing the old footings (mostly of Norman age) in 2 metres thickness of concrete post-tensioned against the old foundation masonry, giving not only an increased bearing area, but shaped to ensure more central loading. Flat jacks were installed to compress the clay, which, together with subsequent pressure grouting, ensured that the new areas of the footings took their share of the load.

The construction of Winchester Cathedral commenced in 1079, with foundations resting on short oak piles driven into the soft marshy ground, some of the piles having been salvaged from the earlier building on the site. In excavating for foundations to support the next stage of building, begun around 1200, the builders came across water 3 metres below the surface and, apparently unable to proceed further with their digging, simply laid beech logs on the base of the excavation and rested the foundations on these. As a result, some 5 metres of compressible clays and peat separated the base of the foundations from the underlying good gravel layer. The inevitable settlement that followed caused the arched vaults to become disturbed, pushing out the walls, and resulted in the building splitting off from the Norman part to the west of it and sliding eastwards, leaving many gaping cracks. Repairs carried out between 1906 and 1911 consisted of grouting the cracks and underpinning the structure. The underpinning was done underwater by an experienced diver William (Diver Bill) Walker, as lowering ground water 5 metres by pumping would have caused further settlements, particularly in the peat layer, under the structure. After excavating a limited length below the foundations down to the gravel, Walker, working in the dark, first placed several layers of jute bags filled with dry concrete on the base of the excavation, which he then split open to allow the concrete to spread and set. With this underwater placement bringing the concrete up to a level adjudged safe to draw the water down to, the hole was pumped out and the remaining concrete to the underside of the foundations placed by normal in-situ methods, before moving on to excavate the next section.

In a structural system as complex as a Gothic cathedral, with designs based on experience and intuition, supported by basic numerical rules of thumb and perhaps some small-scale models, a considerable number of failures must have occurred in the structures and mostly gone unrecorded. As structures are often at their most vulnerable during construction, and their stability strongly influenced by the sequence of construction, most failures would have occurred and been remedied before completion and thus attracted little attention. This attitude is not unknown in construction today, as it is not necessarily in the interests of any of the involved parties – designer, constructor or client – to advertise untoward events. Failures have always constituted part of the learning process in civil engineering, as indeed in most professions, and must certainly have done so in the construction of medieval cathedrals, the knowledge gained becoming closely guarded within the Masonic lodges. Recorded instances of failures show that the medieval builders had an advantage over modern builders in being able to explain the incident by divine intervention. When the central tower of Winchester Cathedral collapsed in 1107, a few years after completion, it was conveniently attributed to the fact that William Rufus had been buried there seven years earlier.

Much of the know-how accumulated by medieval builders had to do with the geometry of members and structure, and with relative dimensions and proportions. While Gothic builders did not know in the mathematical sense that the resultant force within a buttress had to be confined within the middle third of its width to avoid tension or high stress concentrations, particularly at foundation level, they learnt from observation and experience that certain relative proportions were needed to stop tilting or even failure of the buttresses under horizontal thrusts from vaults or flying buttresses. Perhaps the more surprising outcome is that they did not go to the other extreme and design columns and buttresses much larger than necessary to ensure safety. It would have done nothing for the appearance of the buildings, but the master-builders might have been able to sleep easier at night. Modern structural theory combined with the most powerful computers would not be able to improve markedly on the design of cathedrals such as Chartres

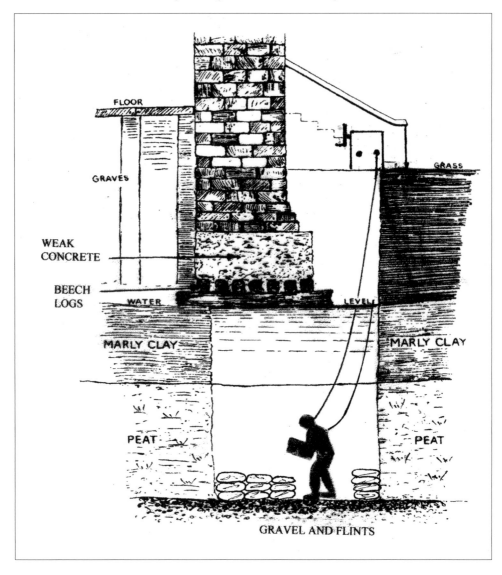

Underpinning of Winchester Cathedral was carried out underwater by William (Diver Bill) Walker between 1906 and 1911. Surface operations, not depicted, included manual operation of the air pump and the control of the length of air line to the diver.

and Amiens, having regard to the materials and construction techniques available to the builders; it would simply remove the human touch or touches (the long periods of construction often meant contributions from several engineer/architects and master builders, and even the combining of different periods, Romanesque with Gothic, Gothic with Renaissance) and result in the charmless uniformity seen in many buildings today. Modern designers have many tools at their disposal, but their initiative can be curbed by the restrictions imposed by codes.

While the naves of Gothic cathedrals achieved soaring heights, emphasised by the geometry of pointed arches and vaults, the widths of naves, commonly 15–16 metres, were modest compared, for example, with the 25 metre span cross-vaults of the Roman Basilica of Maxentius, completed in 312 AD. The attainment of unprecedented vault heights became an obsession with French cathedral builders, each trying to surpass the efforts of earlier compatriots, a race that English builders wisely ignored. Prior to the commencement of Beauvais Cathedral in France in 1225, nave heights had already reached 38 metres at the springing level of the vaults at Rheims Cathedral, begun in 1211, and nearly 43 metres in Amiens Cathedral. These cathedrals represented the challenge that the builders of Beauvais had before them and they met it by designing a structure with the unprecedented nave height of 48 metres. Only Cologne, begun in 1248 and completed in the nineteenth century, matched this.

In contrast with those earlier cathedrals, which apart with some trouble with flying buttresses at Amiens remained stable once completed, Beauvais suffered two major failures and the nave was never completed, the structure as it stands today consisting only of the apse, choir and transepts. Furthermore, the first collapse of the choir vault occurred in 1284, twelve years after the apse and choir had been completed, thus confounding the expectation and normal behaviour that if a masonry structure of this type stands firm on the removal of temporary supports and centring, then it will remain firm indefinitely in the absence of untoward events such as large foundation movements or earthquakes.

Civil engineering has had many examples of failures – from the Meidum Pyramid in Egypt to the Tacoma Narrows Bridge in the USA – resulting from taking a step too far, bred of over-confidence or perhaps carelessness after the successful exploitation of new structural types, sizes, techniques or materials. Unfortunately, no contemporary accounts survive of the collapse of the choir nave at Beauvais, but the obsession to achieve greater height must surely have contributed. A review of attempted explanations of the collapse has been made by Heyman, from which he concludes that it probably started with a trivial localised accident and then spread to the whole of the fabric. This view supports the explanation offered by the nineteenth-century French architect Viollet-le-Duc that the failure occurred initially at the point where one of the lower buttresses met the nave pier (the tas-de-change), each of which actually consisted of substantial masonry-block piers with a parallel system of slender columns, helping to support a masonry block at the tas-de-change; this block, in turn, supported a heavy statue occupying the height between where the upper and lower flying buttresses met the pier.

Viollet-le-Duc considered that shrinkage of the moisture in the joints between the masonry blocks of the main part of the pier threw additional loading into the slender columns, which then fractured, removing support from the masonry block at the tas-de

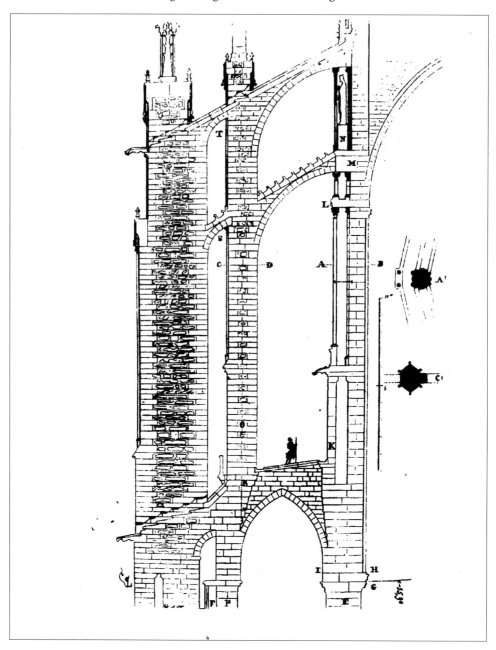

Section through the portion of Beauvais Cathedral that failed, thought by Viollet-le-Duc to have been caused by shrinkage of mortar in pier B, throwing more weight into twin columns A, which then fractured, leading to a progressive collapse.

change, which then slid out. Heyman considers it more likely that the block tilted under the weight of the statue, so driving the line of thrust outside the section of the flying buttress, causing it to collapse. Progressive collapse would then have occurred through the structure as failed sections shed load into adjacent members, perhaps already near their critical loading, thus causing them to fail; this progressive action illustrates all too well the interdependence of structural elements in the Gothic structural system.

The explanations by Viollet-le-Duc and Heyman are consistent with failure occurring twelve years after completion, as the mortar in the joints set very slowly. The unprecedented height, particularly of the clerestory wall, also meant many horizontal joints between the pier blocks, and consequently substantial vertical shrinkage.

By 1337 the choir had been rebuilt with double the number of buttresses, but the Hundred Years War between France and England, which lasted intermittently until 1453, brought activity on the cathedral to a stop. For a time the English occupied Beauvais. By 1500 a new national spirit had emerged in France and work resumed on Beauvais Cathedral transept, which saw completion in 1548. Before starting work on the nave, it was decided to build a masonry tower over the crossing. With their enthusiasm in no way diminished by the collapse of the choir, which in any case had occurred nearly 300 years before, the builders decided to construct a tower rising to the unprecedented height of 155 metres, well outside the limits of their experience and expertise. It compared with 142 metres for the maximum cathedral tower height still standing in Europe at Strasbourg and 123 metres for the spire of Salisbury Cathedral, the highest in England. Built between 1564 and 1569, the Beauvais tower collapsed in 1573.

Two years after completion of the tower, it was found that the two main crossing piers on the choir side were 50 mm and 100 mm out of plumb while those on the nave side, lacking the support of a non-existent nave, were out of plumb by 125 mm and 275 mm. The masons recommended immediate commencement of work on the two nave bays, as well as strengthening of the pier foundations and, as a temporary measure, the construction of walls between the crossing piers. However, procrastination won the day and the chapter delayed two years before putting any work in hand. The tower collapsed thirteen days later, on Ascension Day, 30 April 1573. As the clergy and the people had just left the cathedral in procession, the chapter saw in this the hand of divine intervention and in 1577 decided that 30 April should henceforth be an annual day of celebration for Beauvais. With no further attempt made to erect a tower or construct the nave, Beauvais became, as it remains today, a choir and transept without a nave.

Unlike the French cathedrals, which usually occupied sites centrally within towns, English Gothic cathedrals mostly formed the focus of a group or Close of monastic buildings, with cloisters and a chapter house, in open surroundings near the edge of the town. Although the master masons and engineer/architects of the earliest Gothic constructions came from France, a progressively distinct English style developed, reflecting architectural differences rather than any English contribution to the basic structural form. English builders never had the same obsession with height that possessed their French counterparts and, as a consequence of this, flying buttresses, if used at all, never dominated the appearance of the structure, as they do in French cathedrals such as Notre Dame in Paris or Bourges.

Not surprisingly, the two English cathedrals that show the strongest French influence are Canterbury (1174–84) and Westminster Abbey (1245–69), both designed by French architects. William of Sens raised his structure at Canterbury on portions of the previous Norman building, largely destroyed by fire in 1170. In 1178, when William suffered severe injuries after falling from scaffolding, an English master mason, William the Englishman, took over the work. A much greater length of choir and nave distinguished Canterbury Cathedral from its French counterparts and became a characteristic of the English structures. The early English Gothic style reached its peak in Salisbury Cathedral, the magnificent spire of which, added in the middle of the fifteenth century and thus some 200 years after the body of the cathedral, gives a striking balance to the structure that would have been lost if it had been added to a French cathedral with its much greater height of nave. As Salisbury Cathedral stands on poor ground, this late addition of the tower inevitably led to differential settlements between the tower and the rest of the structure, resulting in visible distortion within the building.

A late and unique development in the architecture of English Gothic was fan vaulting, first introduced extensively in the cloisters of Gloucester Cathedral in 1407 and reaching its full glory in Kings College Chapel in Cambridge (1446–1515). Wells Cathedral boasts the most unusual feature, consisting of inverted arches erected between the pairs of crossing pillars in 1338 for the purpose of providing them with lateral support to arrest bowing observed when heightening of the tower imposed additional loading on them. This bowing effect was exacerbated by bending in the piers induced by the thrusts from the crossing and nave vaults acting on them at different levels.

Salisbury Cathedral.

Crossing pillars in Salisbury Cathedral buckling under the weight of the tower and spire.

Although taking up the Gothic style with some enthusiasm, builders on the northern European mainland contributed very little structurally, with the exception of often substituting brick for stone. In Cologne, Germany has one of the largest of all Gothic cathedrals. It is largely a copy of Amiens Cathedral, but with double aisles to the nave. Work on Cologne Cathedral began in 1248, but progressed very slowly and by 1322 only the choir had been completed; it then effectively stopped until the middle of the nineteenth century, when the nave, aisles and transepts were completed to the original design.

Although the Gothic style had its proponents in Italy, they were more than matched by the Romanesque traditionalists, particularly in Florence. In some measure this may have been a reaction by the Guelph factions who controlled Florence against the Ghibbeline Viscontis who ruled in Milan. Possibly because of a strong German influence in Milan, it built its cathedral in the Gothic style. In 1294, when he started work on the new Florence Cathedral at Santa Maria del Fiore to replace a demolished seventh-century structure, the designer Arnolfo di Cambio certainly envisaged a Gothic structure, but, unlike French Gothic, with a dome over the crossing. Work progressed slowly and stopped at the time of the Black Death. When it resumed in 1350 the two capomaesti then put in charge, Francesco Talenti and Giovanni Ghini, favoured a structure of Gothic form with a vaulted nave, pointed longitudinal arches over the nave piers and external buttresses. A group of artists asked to report on the design opposed the Gothic form in general and external buttresses in particular, and with the backing of the citizens their view prevailed, at least with respect to the buttresses, using instead unsightly tie-rods spanning across the nave to take lateral thrusts. The remainder of the Duomo up to the springing line of Brunelleschi's great dome represented something of a compromise, with its vaulted nave and pointed arches flatter and wider than pure Gothic and with pointed windows hinting at Gothic in the aisles, but decidedly Florentine Romanesque round windows above the aisles and in the tambour or drum supporting the dome.

When the Milanese started construction of their cathedral towards the end of the fourteenth century they opted for the Gothic style, but the period of High Gothic had ended a hundred years earlier and their building illuminates the way in which architectural detail had overtaken structural purity. Its external appearance is a fantasy of white shining marble bedecked with 135 intricate pinnacles and a surfeit of ornamental sculptures. Its double-aisles interior, a little reminiscent of Bourges, with a domed cross-over and fifty-two giant pillars, is dark and gloomy, but able to accommodate a congregation of 40,000, making it the second largest in Europe after Seville. Both the plans and a structural model for the cathedral had been finished by 1387, but these triggered much debate, most of which simply served to illustrate that many of the excellent rules well known to the builders of the twelfth century had been forgotten. In an effort to resolve the matter, the Milanese consulted leading architects from Italy, Germany and France. Much of the debate concerned proportions, which was settled early on in favour of a series of expanding triangles encompassing the cross-section through the nave. When John Mignot arrived from Paris in 1399, at a critical point of the construction, his fifty-four criticisms ranged from structural to aesthetic, worrying officials sufficiently for them to appoint him as cathedral architect only to dismiss him shortly after in 1401. Work proceeded slowly, with completion of the nave and dome by the early sixteenth century and finally the completion of the spires and façade in the nineteenth century.

At the end of the nineteenth century large cracks and fissures appeared in the pillars of the nave and the tyburium, causing great concern for its safety. It was found that each pillar consisted of a core of granitic rock with a circling ring of Candoglio marble, the latter having an elastic modulus five times that of the core and thus carrying most of the pillar loading. Eccentric loading on the pillars from the lantern added to the problem.

Another more recent factor, ground water lowering, has caused further damage. The building stands on foundations placed about 8 metres below ground level, at the then ground water level, the soil profile below foundation level consisting of sands and gravels, with a small amount of silt, to a depth of about 100 metres. Water extraction from these sands and gravels since 1920 has caused large surface subsidence in Milan and, specifically in the cathedral, differential settlements between pillars occurring at a rate of some 5 mm per year. When the possibility of collapse became evident in 1969, reductions in ground water withdrawal were immediately ordered, which resulted in a slow rise in the groundwater table. During 1981–4, damaged portions of pillars were replaced with new competent marble.

10

Heavenly Domes

Hagia Sophia, Constantinople

The inspiration for the magnificent domes gracing Islamic mosques and Renaissance cathedrals came largely from the Pantheon in Rome, built, or certainly extensively rebuilt, by order of Hadrian around 120 AD, and from the Haghia Sophia in Constantinople, built as a Christian church by order of Emperor Justinian and completed in 537 AD. The dome of the Hagia Sophia could hardly be different from the Pantheon: it is much thinner and structurally more sophisticated than the Roman structure, the structural materials are brick and mortar rather than concrete, and it is set on a square base rather than on a round drum.

Early in its existence, Constantinople had been divided into four administrative zones known as 'demes', each named by a colour – red, white, green and blue. These merged eventually into two basic groups – Blues and Greens – each having distinctive civil responsibilities, for example in the construction and maintenance of the infrastructure of the city. Considerable rivalry existed between the two factions, which the Emperors often fostered to their advantage, but the rivalry manifested itself mostly at the hippodrome where the races, four chariots at a time over seven laps, produced savagely contested spectacles, much to the enjoyment of the crowd. It was when the Blues and Greens combined to pursue a common purpose that the Emperor needed to watch himself and in some cases he had to exercise considerable diplomatic skills. One such occasion led to a glorious outcome for the Byzantines and all of humanity, although giving the Emperor Justinian some uncomfortable moments – the commissioning and construction of The Great Church, the Hagia Sophia: the church of Divine Wisdom.

When Justinian imposed swingeing taxes to finance his ambitious military campaigns, the Blues and Greens temporarily forgot their rivalries in AD 532 and combined their efforts to thwart the measure. In the so-called Nika riots that followed, the church occupying part of the site of the present Hagia Sophia, with the same name, was burnt down. The riots almost brought about the abdication of the despairing Justinian; but thanks to the iron resolve of his wife, the empress Theodora, and the loyalty of his army general Belisarius, he survived to such resounding effect that he was able to replace the gutted church with the magnificent structure which has occupied the site ever since, although not without its own traumas.

The Hagia Sophia burnt in the Nika riots had itself been preceded on the site by a church of the same name built in AD 360. This church suffered the same fate as its successor, having been burnt down by mobs protesting against the banishment from Constantinople of the Patriarch, who had dared to criticise the Empress. Even if Hagia Sophia II had not been burnt down, it is certain that Justinian would have commissioned a new building as an imposing and lasting monument to the glory of God and as a reminder to posterity of his own not inconsiderable achievements. Once he had recovered from the shock of the Nika riots, the burning down of Hagia Sophia II must have seemed to him like a final approval from above to go ahead, particularly as this site already had a history of 200 years' ecclesiastical usage. Evincing a sense of history when he entered the newly completed church at its inauguration ceremony in 537, hand in hand with the Patriarch, he exclaimed with some justification, 'Solomon I have surpassed thee.' Although at times he had expressed some limited criticisms of Justinian in previous writings, Procopius, on this occasion, did not spare pen or parchment in fulsome praise of the great structure and the Emperor for his close involvement in its design and construction. Writing of the dome he says 'it seems somehow to float in the air with no firm basis and again it seems....to cover the space with its golden dome from heaven...'

Justinian engaged two Greek master builders, Anthemius of Tralles and Isodorus of Milesia, to design his great church. In his references to these men, Procopius uses the term *Mechanopios* or *Mechanikos*, which translates best as 'engineer' in modern terms. Between them they had written a number of treatises on mathematics, geometry and mechanical devices. It is also clear that between them they combined the skills of architects, structural engineers and surveyors. It is likely that they first constructed the small church of St Sergius and St Bacchus (now a mosque sited between the Hippodrome and the Sea of Marmara). Known as the Little Hagia Sophia, it seems to have been used as an experiment to determine how to erect a dome over a rectangular support system.

Justinian put Anthemius and Isodorus under pressure from the outset to complete the building quickly. Before any work could begin, houses surrounding the burnt-out Hagia Sophia II had to be appropriated to accommodate this much larger structure. As rock underlay the site, foundations presented no problem; but lack of labour for such an unprecedented structure forced the Emperor to scour the known world for skilled workers, the total workforce at the peak of construction reaching some 1,000 foremen and 10,000 other workers. Building materials, too, came from many parts of the Empire. While the walls, dome, semi-domes, arches and vaults are of brick, probably fired locally, piers and lower portions of buttresses are of dressed limestone and granite blocks, and freestanding columns are marble monoliths. Cornices, too, are of marble. Justinian ordered his Governors throughout the Empire to send the most beautiful pieces of ancient ruins in their regions: pillars, marbles and coloured stones came from ancient towns in Anatolia, white marble from Marmara Island, green porphyry from Egriboz Island, yellow marbles from Africa.

Far from 'floating in the air', the dome is supported by four massive stone piers occupying the corners of a square. The transition from the circular dome to the square support is achieved by a spherical triangle, or pendentive, at each corner rising from the supporting

An interior view of the Hagia Sophia.

The Hagia Sophia was built in Constantinople as a Christian church in the time of the Emperor Justinian, and completed in AD 537. Large quantities of brick were used in its construction, including the 31 metre diameter dome, as well as special stone blocks imported from various parts of the Byzantine Empire. The dome has survived a number of partial collapses caused by earthquakes.

column to the coping line around the base of the dome. Although the dome itself has a diameter equal to the sides of the square, the pedentives can in effect be visualised as formed by the spherical surface of a dome having a diameter equal to the diagonal of the square, with the overhangs cut off by vertical planes, leaving four arched openings, surmounted by a dome having a diameter equal to the sides of the square. The arched openings, springing from the support piers, form the curved edges of the pendentives.

Load-carrying arches springing from the pier supports, thence following the curves of the openings, give an efficient means for transferring the vertical loads from the dome and the pendentives to the piers and through them to the foundations; but these arches do not provide an effective means of resisting the outward lateral thrusts from the dome and pendentives. Along the east–west axis of the Hagia Sophia, large semi-domes with the same diameter as the dome counter the thrusts and, at the same time, extend the open, unencumbered length of nave in this direction, making it rectangular rather than square in plan. The semi-domes themselves exert outward thrusts and these are absorbed by two small semi-domes on the flanks of each main semi-dome, and by barrel vaults (which are inherently stiff in the direction of their alignment) along the axis, central between each pair of small semi-domes. The barrel vaults at each end, and the small semi-domes, in turn distribute the thrusts into the vaults, arches, walls and piers making up the remainder of the structure, including the narthex (vestibule) at the western end and a small apse at the eastern end.

Unlike the highly efficient system described above to absorb the east–west thrusts, the structural system absorbing the north–south thrusts has serious inherent weaknesses. No provision is made to counter the outward thrusts on the main arches, which span 31.6 metres, other than secondary arches, below the main arches, but with greater width and springing from projections of the main piers, giving them a lesser span of 22.6 metres. Not only does this create lateral bending and shear stresses in these arches, which they cannot properly resist and in consequence bow outwards, but lateral loads are transferred through the arches to the piers. This concentration of the north–south lateral loads at the tops of the support piers contrasts with the much more even distribution of east–west lateral thrust into the body of the structure through the system of large and small semi-domes and barrel vaults. Although these north–south concentrated forces eventually become dispersed through the system of buttress walls above gallery roof level connecting to the outer buttress piers, and through vaults and arches spanning the galleries and aisles flanking the northern and southern sides of the nave, this is accomplished at the expense of considerable distortions and shearing in the masonry.

Measurements by Emerson and Nice (Dunbarton Oaks Centre for Byzantine Studies, Harvard University) showed the extent of movements and distortions caused by the lateral thrusts from the dome. In particular, they emphasised the inability of the structure to give full support in the north–south direction, as the piers have tilted outwards three times as much to the north and south compared to the tilts to the east and west. As a result of these movements the dome, originally circular in plan with a diameter of about 31 metres, now measures 31.85 metres along its north–south axis and 30.85 m along its east–west axis. These figures show that as the dome spread in the north–south direction, the semi-domes on the east and west sides actually exerted a squeezing effect, slightly reducing the width of the dome; this is also seen in vertical projections of the main supporting arches, those along the north and south sides showing substantial outward bowing and those along the east and west sides showing slight inward bowing. This also underscores the effectiveness of semi-domes in providing lateral support.

Paradoxically, the partial collapses suffered by the dome in the sixth, tenth and fourteenth centuries all occurred in the east and west quadrants, because the large north–south tilting of the piers spread the bases of the east and west arches causing their crowns to drop, with the inevitable collapse of the adjacent portion of the dome and, consequently, the semi-domes. The collapses all occurred during or following earthquakes, but these were a trigger for the collapses, not the cause, which was the inherent weakness of the structure. After the collapse of the eastern arch in 557, reconstruction was carried out by Isodorus (nephew of the original master-builder of the same name), who raised the height of the dome by 6.3 metres to reduce lateral thrusts. On completion of the work in 562, Justinian reconsecrated the church. In the tenth century the western arch collapsed and in the fourteenth century it was again the turn of the eastern arch. In each case portions of the adjacent dome and semi-domes again collapsed and had to be reconstructed.

Most of the bricks making up the bulk of the structure, and fired locally, measured 0.375 metres square and, on average, about 45 mm thick. As they were commonly separated by mortar joints 50 mm to 60 mm thick, and thus thicker than the bricks themselves, the mortar became the most important element determining the load-carrying characteristics of much of the structure. Some of the considerable distortions suffered by the building, particularly in its earlier years, may be attributable to this predominance of a slow setting mortar, made up of slaked lime, sand and crushed brick or tiles, in walls and buttresses subjected to large thrusts from the roofing. Larger imported bricks, up to 0.7 metres square, were used for specific parts of the building, notably in the main north–south and east–west arches supporting the dome, which consisted originally, or certainly after the 562 reconstruction, of double rings of radial bricks of the largest size. As a result of later reconstruction the western arch now has a greater thickness of about 2.8 metres and a width of 4 metres compared to 3.5 metres for the original arches. Procopius gives an instructive account relating to the erection of one of these main load-carrying arches:

> One of the arches which I just now mentioned, the one which stands towards the east, had already been built up from either side, but it had not yet been wholly completed in the middle, and was still waiting. And the piers [pessoi], above which the structure was being built, unable to carry the mass which bore down on them, somehow or other suddenly began to crack, and they seemed on the point of collapsing. So Anthemius and Isodorus, terrified at what had happened, carried the matter to the Emperor, having come to have no hope in their technical skill. And straightaway the Emperor, impelled by I know not what, but I suppose by God (for he is not himself a master-builder), commanded them to carry the curve of this arch to its final completion. 'For when it rests upon itself,' he said, 'it will no longer need the props [pessoi] beneath it.'

In this passage the term 'pessoi' can only relate to the timber centring supporting the partly completed arch (and not the piers, as translated) and Justinian's assurance that, when completed, the arch would 'rest upon itself' and be self-supporting showed a commendable understanding of the behaviour of a radial arch, whether by the Emperor or by God. Anthemius and Isodorus must also have been fully aware of the structural characteristics of an arch and perhaps decided to involve the Emperor in the decision as an insurance against the possibility of some untoward behaviour outside anything they had previously experienced. It is equally possible that despite protestation to the contrary, Procopius is flattering the Emperor, possibly in an attempt to mend some bridges, having been critical of Justinian in earlier writings.

The dome itself consists structurally of forty radial brick ribs, 1.2 metres wide and 2 metres deep where they rise from the cornice of marble blocks, cramped together, separating the dome from the pendentives. At this lower level windows separate the ribs which, externally, rise vertically from the cornice to the tops of the windows, where they cut back to meet the dome, giving the impression of a supporting drum. Above the windows 0.7 metre thick webs separate the ribs, which gradually become less prominent approaching the crown, giving a smooth interior surface at the top of the dome. The ribs and webs are thicker in the western portion of the dome following reconstruction. The semi-domes have a thickness of about 0.8 metres and probably consist of two thicknesses of the ordinary bricks. Lead sheeting covers the exterior of the dome and semi-domes.

When Sultan Mehmet conquered Constantinople in 1453, thus ending over a thousand years of Byzantine civilisation, he converted the great church, now in a state of poor repair, into a mosque. He also established a substantial fund to meet its continuing maintenance costs. His first expenditure was the erection of a minaret, two of the other three minarets now gracing the structure being the work of Sinan, who also carried out other major restoration, including the reinforcement of existing supports and the provision of additional supporting walls. Further major restoration carried out in 1847–9 by the Swiss engineer Fossati included straightening some of the columns and reinforcing the dome with iron rings. The opportunity was taken at the time to make detailed drawings of the structure, based on measurements taken by Fossati. Additional work to strengthen the structure undertaken in 1926 included the addition of another iron ring to the dome and replacement of the lead sheeting on the dome and semi-domes. In 1935 Atatürk had the building opened to the public as a museum.

The gold figural mosaics originally adorning the interior of the church did not survive the iconoclastic period of 726–846 and the figural mosaics seen today date from the ninth and tenth centuries. The Ottomans covered these mosaics with whitewash, which Fossati removed during his reconstruction work in the nineteenth century, only to re-cover them with new plaster, presumably at the order of Mustafa Resit Pasa, the Grand Vizier. An earthquake in 1894, which caused widespread damage in Istanbul, brought down plaster in the Hagia Sophia and damaged some of the mosaics. With the building now a museum, the mosaics are once again fully restored and on show.

Ottoman Mosques

Although built originally as a Christian church, the Hagia Sophia (renamed the Aya Sofiya by the Ottomans) served both as a structural example and a challenge to the builders of the great mosques of the Ottoman Period, none more so than Mimar Sinan Agha, one of the most outstanding and prolific engineer/architects of all time. When Süleyman returned from his Balkan campaign in 1543 to learn that his son of 22 and intended heir, Seyzade Mehmet, had died, he ordered Sinan to build a mosque in memory of his favourite son. He built a mosque with a central dome, and no doubt mindful of the shortcomings of the Hagia Sophia design, supported the 18 metre diameter dome laterally with four symmetrically placed half-domes, each broadening out into two exedras (semi-circular roofed spaces). Four small domes fill in the corners of the rectangular plan shape and four turrets projecting above the walls, one in each corner, give added weight and stability to the structure. The four great arches marrying the main dome to its buttressing semi-domes spring from massive octagonal fluted piers.

Enriched with booty from his campaigns in the Mediterranean, the Middle East and Europe, Süleyman gave Sinan the order to build for him a mosque, the Suleymaniye, on a prominent site overlooking the Golden Horn. In an apparent effort to avoid the monotony of a totally symmetrical design, as in the Sehzahde mosque, Sinan produced a design more akin to the Hagia Sophia, semi-domes on two of the opposite sides (in this case the north-west and south-east faces) taking the thrusts from the dome and linked at the corners to exedras to further distributing the thrusts, while on the other two sides the load-carrying arches are exposed and unsupported laterally, except at their springings, where massive buttress walls projecting outward from the arch supports take the thrusts. These are more integral with the structure of the building than in the Hagia Sophia. Fenestrated curtain walls filling the main arch spaces in the north-east and south-west sides are supported by a large central arch and two flanking smaller arches, the ends of the central arch and the inner ends of the outer arches resting on 10 metre high piers of Egyptian granite. The four main arches supporting the 26.5 metre diameter central dome rest on four massive limestone piers. The dome soars to a height of 53 metres at its crown, giving a height to diameter ratio of 2:1, one of the simple ratios, along with 3:1 and 3:2, commonly used by Sinan in his structures. The naves flanking the north-east and south-west sides of the domed interior are each topped by five domes, three larger ones and two smaller ones. All domes and semi-domes are constructed of light-weight fired brick.

Sinan himself regarded his Seyzade mosque as the work of an apprentice and the Sulimaniye as his mature work. But he personally rated above these, as his masterpiece, the Seliminiye at Edirné, built for Sultan Selim II, known as Selim the Sot for his debauched lifestyle. Having moved his family to Edirné in 1566, Sinan was in a position to supervise personally the construction, which commenced in 1569 and reached completion six years later in 1575, Sinan by now having reached 86 years of age. Selim ordered the construction of this mosque to celebrate the capture of Cyprus by his troops, but a few months later the defeat of his navy at Lepanto effectively put paid to the Ottoman mastery of the Mediterranean. His reason for building the mosque in Edirné can only be guessed at, but it may have been because he could not identify a sufficiently prominent site in Istanbul to satisfy his ego, whereas the hilltop site he chose in the summer capital of Edirné dominates the surrounding area. Perhaps the fact that Sinan had moved his family there a few years before might also have influenced him.

The 31.3 metre diameter of the dome matches that of the Hagia Sophia, as does its height of 43.5 metres from fountain to crown. Departing from the pattern of four massive piers in a square pattern supporting the dome, which he had favoured previously in line with earlier mosque builders, Sinan adopted in Edirné an octagonal pattern of eight piers of granite and marble. Furthermore, the piers are set into the walls, giving an uncluttered and spatially sophisticated interior. The square of the exterior plan is completed by four small semi-domes in the diagonal corners, but the lateral thrusts from the dome are absorbed by turretted buttress walls integrated into the structure of the building, the elegance of which is heightened by its having, at 71 metres, the tallest minarets of any mosque in Islam.

In the mosque of Sultan Ahmet, built in Istanbul close to the Hippodrome between 1609 and 1617 – or the Blue Mosque as it is popularly called from the predominant colour of its tiled interior – the Imperial Architect, Mehmet Aga, reverted to the Seyzade model, providing semi-domes on all four sides to take the thrusts, even though it had a smaller dome, 23.5 metres diameter, than either the Hagia Sophia or the Suleymaniye. Pursuing the building of his great mosque with the same single-mindedness that he pursued his pleasures, Sultan Ahmet had several palaces pulled down to clear a site fitting for his great structure and monopolised the supply of stone, marble and tiles during its construction, to the detriment of other important works.

The Hagia Sophia also set a challenge in the rapidity of its construction time, fractionally less than six years; but the Ottoman builders matched it, the Suleymaniye taking six years (1550–6) and the Blue Mosque eight years (1609–17). These times are remarkably brief compared with the much slower construction of the great Gothic cathedrals, Chartres almost setting a record for brevity with its thirty years and many of them taking up to 100 years or more.

Ottoman builders made much greater use of stone in mosques than used in the Hagia Sophia, and they made considerable use of iron ties between arch springings to absorb the thrusts, even installing them between the springings of semi-domes. In the interior of the mosques the structural forms are more prominent than in the Hagia Sophia, not least exemplified by the massive piers or 'elephant's feet' supporting the dome of the Blue Mosque, and there is less emphasis on spatial separation between the central nave and peripheral aisles.

Santa Maria del Fiore

When Filippo Brunelleschi started work on the dome of the cathedral of Santa Maria del Fiore in 1420, he had two major constraints to contend with. Firstly, the supporting drum or tambour had already been completed to an octagonal shape in plan, compelling him either to continue with this shape for the dome or to adopt a circular shape with pendentives. Furthermore, the tambour had no external Gothic-style buttressing to absorb lateral thrusts and no internal cross-ties either, such as those that mar the internal appearance of the nave of the church. Consequently, lateral thrusts had to be absorbed within the dome itself, a task more difficult to accomplish with the octagonal shape than it would have been with a circular shape. Nevertheless, Brunelleschi opted for the former, producing a dome of striking appearance. Brunelleschi imposed upon himself the second constraint, having won the commission to build the dome on the claim that he could do so without the need for centreing that would have required vast amounts of timber, costing enormous sums of money both to supply and erect. It was largely this problem that had delayed construction of the dome for some decades. In making the claim that he could eliminate the need for centring, he showed remarkable insight into the structural action of a dome. There is little doubt that he had been dwelling on the manner of construction of a great dome for this cathedral in his beloved Florence for many years, certainly since his return in 1404 from Rome, where he had taken the opportunity to study the dome of the Pantheon, which was of almost equal span. His ideas had matured by 1420 and consequently construction proceeded to completion in 1436, almost without a hitch and with little change to his own initial specifications. Since its completion it has dominated the skyline of Florence in a majestic and compelling way, hardly matched by any other man-made structure in the world, and it is a testimony to what can be achieved when civil engineers and architects work together in harmony; in this particular case these two disciplines resided within a single individual, the separation of the two professions not occurring until the nineteenth century. Other outstanding achievements of the Renaissance, not excluding Leonardo's *Mona Lisa* and Michelangelo's *David*, almost fade into insignificance against Brunelleschi's magnificent creation.

Construction of the cathedral began in 1296 with a nave of essentially Gothic design, but even this early a domed roof over the crossing of the nave and transepts appears to have been envisaged. Most of the body of the present church, however, dates from the second half of the fourteenth century, during which period intense disputes arose in Florence between Gothic and Romanesque factions and, in particular, whether or not the dome should be encumbered with Gothic-style external buttresses. Both sides seem to have accepted that internal visible tie bars should not be used, as they had been in the nave vaults to try and halt cracking. A Commission set up by the city to resolve the disputes had, as one of its members, Ser Brunelleschi, later to become father of Filippo. The Commission decided against external buttressing.

A large brick model, complete with dome, built adjacent to the site served as a frame of reference throughout the construction period from 1367 to the end of the century. Filippo, born in 1379, no doubt studied the model and with the encouragement of his father, a lawyer and notary, watched critically the progress of the construction work. By 1400 the four piers to support the dome had been completed and the principal pointed arches connecting them were under construction, so making irreversible the decision to have an octagonal drum. Around 1400 a new architect, Giovanni d'Ambrogio, who favoured the Gothic style, took over the work and proceeded to design tall buttressing structures making up the tribunes, topped with small semi-domes and containing the chapels at the opposite end of the church to the nave. A new Board of Advisers, including Ghiberti and Brunelleschi, forced Giovanni to lower these supporting structures.

Brunelleschi's magnificent dome of the Santa Maria del Fiore dominates the skyline of Florence.

The dome of the S. Maria del Fiore stands on an octagonal drum which, together with the contrast between the red tiles and exposed white marble ribs, gives it its striking appearance.

It seems certain that at this time Brunelleschi must already have been entertaining thoughts of building the great cupola for the cathedral, but having lost a competition to Ghiberti in 1401 to design the great bronze doors of the baptistery of San Giovanni, he took himself off to Rome with his friend Donatello to study Roman buildings and, in particular, the Pantheon. He made extensive drawings of the many structures he studied. He returned to Florence in 1407 in time to attend a congress called to discuss the raising of the cupola, but having made sufficiently shrewd comments to indicate his clear understanding of the problem, he decided to play hard to get and returned to Rome. The anticipated invitation from the wardens of the Santa Maria del Fiore and the consuls of the Wool Guild to return to Florence duly arrived and he graciously allowed himself to be enticed back. On this occasion he presented his ideas in a little more detail, while stressing the great difficulty of constructing an octagonal dome compared to a circular dome. As time went by without his being given the commission to build the cupola, he became impatient and refused to build a model. Instead, he recommended that architects and engineers from Italy, France, Germany England and Spain should be invited to assemble in Florence and pool their ideas on how to raise the cupola. Filippo's motives in this may well have been that it would serve to demonstrate his own superior understanding of the problem and, secure in this belief, he took himself off once again to Rome in 1417 despite the offer of a grant from the wardens to keep him in Florence.

In the meantime, from about 1410, work on the tambour went on under Giovanni d'Ambrogio, and it is possible that Brunelleschi advised on the insertion of its large round windows. It was built free of external buttressing. Clearly this 10 metre high drum, completed in 1413, with its octagonal shape and inability to resist substantial thrusts became a major factor in the design of the dome. It also determined its dimensions, the circle just touching the faces of the octagon having a diameter of 42 metres, and that through the corners a diameter of 45.5 metres, giving a mean diameter of 43.7 metres. This compares with 43.5 metres for the Pantheon and 41 metres for St Peter's in Rome, for which Brunelleschi's dome was in some measure the inspiration. The Hagia Sophia dome is only about three-quarters the diameter of Santa Maria del Fiore. There can be no doubt that the width of the Florence dome was deliberately chosen to rival that of the Pantheon, the construction of which, however, could not have been more different, its perfectly hemispherical concrete dome springing from a massive concrete drum 6 metres thick. Brunelleschi lessened the weight of the dome towards its top by reducing the thickness of the double shells, whereas the Roman builders used travertine and tufa as aggregate in the base of the concrete shell, changing to lighter tufa and brick with increasing height and much lighter pumice at the highest levels.

The year 1418 saw the assembling in Florence of the international masters as proposed by Brunelleschi, and the church superintendents invited them to submit designs for the dome, these submissions to include methods of construction. When the superintendents initially rejected his claim to erect the dome without centring, Brunelleschi built a model and wrote out a careful account of how he intended to proceed. The superintendents compounded and, mindful of the enormous savings to be made by not using centring, they entrusted the work to him, but at the same time appointed Ghiberti as his associate.

Apart from small changes in detail during construction, the completed dome conformed closely to the design put forward by Brunelleschi in 1418. Up to a height of about 3 metres the dome consists of solid sandstone 4.3 metres thick, above which it becomes a double shell, the inner one having a thickness of 2.2 metres at its springing, reducing to 2.0 metres at the oculus supporting the lantern, and the thinner outer shell having a thickness of 1.0 metre at its springing, reducing to 0.4 metres at the oculus. The space between them increases slightly from 1.2 metres at the springing to 1.4 metres at the top. The outer shell gives weather protection to the inner shell and rib system that provides the structural strength and, at the same time, by covering this structural system, enhances the external appearance. The vitality and commanding appearance of the dome derives in part from its size and geometry, and in part from the contrast between the red tiles coating the face of the outer shell and the exposed white marble over the main corner ribs. A section through the full width of the dome has the shape of a pointed arch, truncated at the top by the oculus supporting the lantern, a shape which, although echoing the despised Gothic style, offered, as Brunelleschi realised, advantages both in its ease of construction and in its structural action.

The shape of the internal shell in vertical section is a circular arc, rising vertically at the springing and meeting the oculus at an angle of 28° to the horizontal. The centre of the circular arc is on the line of the springing, with a radius equal to slightly greater than five-sixths of the width of the tambour. Not only is this shape aesthetically pleasing, but it provided Brunelleschi with an optimum form both in the design of the cupola and in its construction.

The shells consist of sandstone up to about 7 metres of their height and brick above this, the bricks having the advantage of being lighter and able to be formed to various shapes as required. The bricks are invariably 50 mm thick, but vary in length, breadth and shape according to requirements. In addition to the eight main corner ribs, which are about 4 metres wide at the foot, there are two ribs, each 2.3 metres wide at the foot, in each of the two sides, separating the outer and inner shells, making twenty-four ribs in all. These are again an unmistakeable Gothic feature, but well hidden. The internal brick ribs are integral with the shells. Seven chains of sandstone blocks, cramped together with lead-coated iron, girdle the dome at three levels, the lowest level, with three chains, coinciding with the top of the solid sandstone portion of the cupola. Transverse or radial sandstone beams, also embedded at this level, are visible on the exterior face of the dome. The remaining four sandstone chains occur in pairs, one pair at about one-third the height of the dome and the other pair towards the top. These sandstone chains are supplemented by iron chains and, at a level intermediate between the two lower sandstone chains, by a wooden chain of oak beams bolted together with iron cover plates. The wooden chain probably contributes little to the hoop strength and may have been included at the insistence of Ghiberti. The inclusion of chains by Brunelleschi in his initial specifications shows that he clearly understood the need to provide substantial hoop strength to counter tensile forces set up by the outward thrusts of the dome, against which the supporting tambour, with no external buttressing and no internal cross-ties, could offer little resistance.

Another major structural feature included by Brunelleschi in his final design consisted of a series of 0.6 metres deep horizontal brick arches encircling the

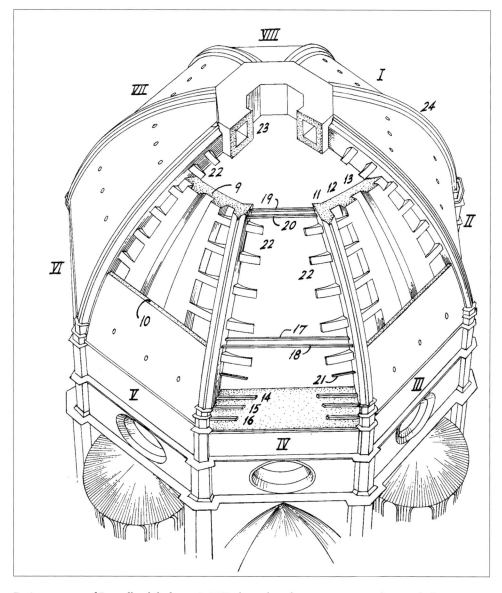

Basic structure of Brunelleschi's dome: I–VIII, the eight sides; 9, 10, inner and outer shells; 11, corner ribs; 12, 13, intermediate ribs; 14–20, stone rings; 21, wood chain; 22, 'small vaults'; 23, key ring; 24, ridges.

dome at nine levels 1.9 metres apart, each set of nine arches spanning one side of the octagon, and thus reducing in span with increasing height. Springing from the corner ribs, the arches, on reaching the intermediate secondary ribs, become incorporated into these, while between the intermediate ribs they disappear from the sight of the observer standing in the space between the shells, as they become entirely contained within the outer shell. It is likely the intent of these arches was to support the outer shell.

The dome terminates at the top with a hollow brick octagonal ring beam, on top of which stone slabs support the lantern. Ambulatories or passageways circle the dome at three levels within the hollow space within the shells, the lowest at the springing of the double shells and the uppermost at about two-thirds of the dome's height. Stairways connect the passageways, and steps resting directly on the inner shell lead from the third level ambulatory to a fourth passageway within the hollow ring beam below the lantern. The outer shell is pierced by circular openings at three levels.

The largest cracks in the dome and tambour have occurred within the sides of the octagon which lie above the four solid pier supports and not above the four arches spanning between the solid piers. This seems paradoxical at first sight and has led to speculation that reinforcement has been placed within the masonry above the arches. This is unlikely. In fact, the pattern of cracking can readily be explained by the structural configuration. If the octagonal annulus of the tambour and lower courses of the dome are opened up to lie in a single vertical plane, the configuration is one of a continuous beam of variable depth over four supports. The greatest depths are over the solid supports and the least depths at the centres of the spans, over the crowns of the arches. A uniform load on such a beam sets up negative moments over the solid supports, thus inducing tensile stresses which, added to tensile stresses caused by the dome action, results in cracking in the higher portions of the tambour and lower portions of the dome. The positive moments set up at mid-span induce compressive stresses in the upper portions of the tambour and tensile stresses are countered by horizontal stresses at the crown set up by the action of the arches themselves.

Some cracking is inevitable in a brick or masonry dome and those in the Santa Maria del Fiore have been carefully mapped and subject to constant monitoring by the Instituto Geografico Militare in collaboration with the Engineering Department of the University of Florence. It is an interesting characteristic of dome construction that when rings of brickwork are placed, compressive stresses are set up between the bricks, as they tend to collapse inwards; but as the height of construction above a certain level increases, the compressive stresses change progressively to tensile stresses as the lower part of the completed portion of the structure tends to burst outwards. As a consequence of this, the radial vertical cracks commonly found in domes tend to start at the springing and extend partway up the dome and downwards into the supporting structure. This is the pattern observed in the dome of the Santa Maria del Fiore.

As the final form of the dome, as far as can be examined at least, coincides fairly closely with Brunelleschi's original specification, he obviously had a clear vision of the dome's structure in his mind from the outset. It was not so much the structure as such, however, that persuaded the authorities to entrust him with the task, but rather his claim to be able to construct it without close-sheeted centring supported by a heavy timber structure or armature. This would have been enormously expensive. He must therefore have had a clear vision also of the construction methods required to justify his claim, which he demonstrated by the construction of large models. It can only be speculated whether or not he drew some confidence from the knowledge that in Egypt large vaulted structures had been constructed for over 2,000 years without centring, using pitched brick construction. It is not unrealistic to compare the two types of construction because in a dome, as in a pitched brick vault, once a ring is complete and the mortar gains some strength, it becomes self-supporting and also provides support for the next ring. In the lower part of a dome the inward inclination of the joints is very small and with carefully prepared mortar there is no tendency for sliding of the bricks along the joints. It differs little from building a stone or brick wall, and needs no lateral support. As the height increases the tendency for the bricks to slip inwards increases as the joint inclinations become steeper, but once a ring has been completed its tendency towards inward slip generates tangential compressive stresses, imparting stability to the ring.

The two principal methods likely to have been employed by Brunelleschi were to construct the rings in self-supporting sections and to provide localised support as necessary within each section. In constructing a dome without the use of centring, compared to erection with centring, the workmen would have placed the bricks from the scaffolding within the dome in the former case and outside the dome in the latter case. Brunelleschi's scaffolding would have consisted of a light timber framework supported on beams thrust into horizontal holes at the base of the dome, and possibly at progressively higher levels, this timber framing in turn supporting a working platform which could be moved upwards in stages as necessary. A herringbone pattern is a distinctly visible feature in the brickwork above the second ambulatory, created by placing bricks vertically on edge at intervals in otherwise horizontal brick courses. Each vertical brick rises through four or five horizontal courses, thus creating a diagonally ascending band. These bricks cannot have been set like this for structural reasons as they weaken the fabric of the brickwork against tangential tensile forces, so it is likely that they served the purpose of marking the ends of construction sections. Once in place, they not only provided set markers against which the intervening horizontal bricks could be positioned, but they also provided stable buttresses to hold these bricks in place. Temporary support for the bricks could have been provided either by workmen holding them in place while the mortar set slightly, or by flying localised supports secured to timbering projecting from the workmen's scaffolding.

Imbued with a tremendous and long nurtured ambition to build the great dome, and with an absolutely clear vision of both its final structure and method of construction, Brunelleschi should have felt unalloyed pleasure and indeed excitement when given the commission to start work. To his considerable chagrin, however, the superintendents appointed Ghiberti as an equal associate with Brunelleschi in the work and with an equal salary. He endured this situation until 1426, but not without expressing his views of Ghiberti in forceful terms and on frequent occasions. Whether or not he could have happily accepted as an equal associate someone with more engineering competence than Ghiberti is a matter for conjecture, but it seems unlikely, despite Vasari's claim that Filippo 'was endowed with outstanding personal qualities, including such a kind

Brickwork in the shells of the dome of S. Maria del Fiore showing patterns of vertical bricks marking the end of construction lengths.

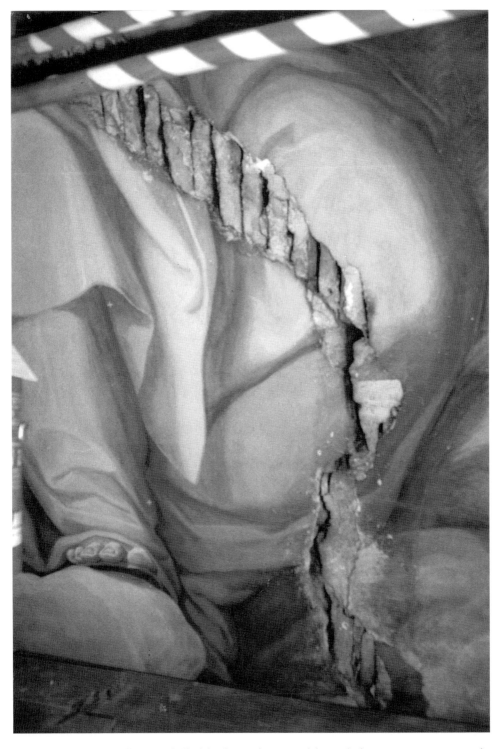

Example of cracking in the inner shell of the dome of S. Maria del Fiore before restoration. Most cracks conformed to a radial pattern starting at the springing, extending partway up the dome and downwards into the tambour.

nature that there was never anyone more gentle or loving'. This gentle, loving nature did not stop him from instituting a little deceit to rid himself of Ghiberti. The dome having reached a point where two specific design decisions had to be made, one concerned with the provision of scaffolding and the other with an encircling chain, Brunelleschi feigned illness and took himself off to bed. As he anticipated, work came to a halt. When the superintendents reprimanded him he expressed surprise that Ghiberti had not been able to issue the necessary instructions for the work to continue. In a show of conciliation, however, he suggested that he should take on one of the design tasks and Ghiberti the other. Ghiberti chose the chain and produced a design that Brunelleschi quickly showed to be inadequate, and replaced it with a much superior design of his own, having already solved the design problem with the scaffolding. At this stage Ghiberti withdrew from the work, leaving Brunelleschi alone to complete the dome in 1436.

St Peter's, Rome

It is said that a total of twenty-three architect/engineers worked on the design and construction of the dome of St Peter's in Rome, including a number of illustrious names, not least Michelangelo Buonarotti. The design changed consequently several times between laying the foundation stone for the new building in 1506 and its completion in 1602. In 1503 Cardinal Giuliano della Rovere became Pope, taking the name Julius II, and immediately set about repairing the damage, as he saw it, done to the existing church by his Borgia predecessors Callixtus III and Alexander VI, and by Alexander's son Cesare. His detestation of the Borgias stemmed in no small measure from the fact that they had continually thwarted his own ambitions. As one of the outward signs of this restoration of the dignity of the Church he commissioned the leading architect of the day, Bramante, to build a great new cathedral to replace the old basilica which had been built in AD 334 on the site of the martyrdom of St Peter. The old structure had in fact been demolished at the order of an earlier pope, Nicholas V, but rebuilding had made little progress by the time Julius II had taken over the pontificate.

Essentially an architect rather than an engineer, Bramante's design for the dome of St Peter's did not match the brilliance of his design for the main body of the church. Circular in plan with a diameter of 41.2 metres and semi-circular in elevation, its solid masonry construction had none of the engineering sophistication of Brunelleschi's creation in Florence and displayed much more the influence of the nearby Pantheon, even to the detail of having a thickened lower portion with a stepped external profile. Fortunately, it was never built. Bramante died in 1514. His immediate assistant, Giacondo of Ponte Notre-Dame fame, died shortly after, to be succeeded by Giuliano da Sangallo; however, Bramante had recommended that he should be succeeded by his nephew Raphael, whose qualities as an architect or engineer in no way matched his exquisite skills as a painter. He died in 1520 at the early age of 37 having made, perhaps fortunately, very little contribution to the cathedral. His successor, Antonio Sangallo, nephew of Giuliano, who remained in office until his death in 1546, redesigned the cupola, retaining Bramante's solid masonry shell, but eliminating the stepped-up rings and changing the elevation to

St Peter's, Rome, with dome designed by Michelangelo and others.

a surface generated by the revolution of a segmental arc. He did not construct the dome, or even its supporting drum, his main achievement being to construct the vaulted nave of the cathedral. He also strengthened the pillars that Bramante had constructed.

As his first action on his appointment as architect/engineer to the works in 1547, Michelangelo sacked the entire workforce set up by Antonio da Sangallo and appointed men he felt he could trust, thereby upsetting a number of influential people in Rome, who tried unsuccessfully to have him removed. This is unlikely to have caused him any sleepless nights as he had an irascible nature, perhaps deliberately cultivated, and had never courted popularity. He also strongly opposed all forms of graft, thus putting him out of resonance with the times. Although he did not build the dome himself, he further strengthened the piers and completed the drum, and changed the design of the dome to a double-shelled brick structure with sixteen longitudinal ribs. He favoured the hemi-spherical form, at least initially, but may later have modified this to the much more efficient Brunelleschi form finally adopted for St Peter's, and previously proposed by Sangallo. He had a number of wooden models built, one of which survives in a form modified by Giacomo della Porta who, with Domenico Fontana, completed the construction of the dome in the five-year period 1585–90, Michelangelo having died at the advanced age of 89 years in 1564. The method of building from internal scaffolding followed closely that pioneered by Brunelleschi.

Brick constituted the predominant material of the cupola as finally constructed, with three iron rings inserted to absorb the horizontal thrust. These, however, proved inadequate and extensive cracking occurred which occasioned, a century and a half after its construction, the first mathematical structural analyses of a major structure. These analyses numbered among several factors marking the transition in civil engineering in the middle of the eighteenth century from an empirically to a scientifically based discipline, but without in any way diminishing the need for the civil engineer to apply experience, exercise judgement, display imagination and to have not a little of the artist in his or her soul.

In 1743 Pope Benedict XIV commissioned three mathematicians, Le Seur, Jaquier and Boscowich, to make an analysis of the dome of St Peter's to ascertain the cause of cracking and devise suitable remedial measures. They employed a technique well enough known to mathematicians at the time and now called the Principle of Virtual Displacements, assuming the cracks to be hinges around which the intact parts of the masonry turned. The work done by the rising or sagging heavy masonry sections was ostensibly equated to the energy of the horizontally expanding iron tie rings, but the erroneous assumption was made that the iron rings exerted a constant force independent of their elongation. They recommended, not surprisingly, the addition of more iron rings.

While agreeing with the proposed remedial measures, the Italian mathematician and engineer Giovanni Poleni, also asked by the Pope to carry out an analysis of the dome in common with a number of others, disagreed strongly with the approach of the three mathematicians and felt that the mechanism they adopted was too theoretical. He based his own analysis on the observation that the cracking tended to rise from the springing towards the crown and so divide the cupola shell into lunes or 'orange slice' segments. For his analysis, he divided it into fifty such lunes, the width of each at the base thus equalling one-fiftieth of the perimeter, diminishing to zero width at the crown. He then considered the stability of arches consisting of opposite pairs of lunes leaning against each other at the crown, applying the proposal put forward by Hooke in 1675: '...as hangs the flexible line, so but inverted will stand the rigid arch'. Poleni considered correctly that if this inverted thrust line lay completely within the masonry then each of the arches, and hence the sliced dome, whether cracked or uncracked, would be stable, provided of course that the lateral thrust could be contained at the springings. He threaded thirty-two beads of varying sizes on a string, each with a weight corresponding to the unit length of the lune it represented, and hung the string from two supports. When he inverted the curvature of the string and superimposed it onto the cross-section of St Peter's dome it lay entirely within the masonry, allowing him to pronounce the dome safe despite the cracking. Measurement of the inward pull of the string at its supports enabled him to estimate the thrusts of his 'orange slice' arches, and consequently the hoop stresses that had to be absorbed within the masonry. This led him to recommend that additional iron rings should be provided to resist these stresses, and he performed strength tests on the iron to determine the number required. Five additional tie rings were added in 1743 and 1734 under the supervision of the architect Vanvitelli.

St Paul's, London

The area on Ludgate Hill in London occupied by St Paul's Cathedral has been the site of a number of Christian buildings since Anglo-Saxon times. A seventh-century church built by Ethelbert, King of Kent, burnt down in 1087, the last year in the reign of William the Conqueror, and its successor started by Bishop Maurice, having suffering damage by fire during the reign of Stephen (1135–54), was completed in the thirteenth century, partly Norman and partly Gothic in construction. The bulk of the construction, using limestone shipped to the site from Caen, was completed during the early Norman period and included a vast gloomy undercroft, a high nave with massive pillars rising to support the vaulted stone roof, north–south transepts and huge piers and arches supporting the 87 metre high central tower, itself erected during the later Gothic period, which also saw the construction of the choir. This immense structure nearly 180 metres in length, the longest Christian church in the world at that time, dominated the city skyline from all points of the compass, not least because of its tapering spire of timber and lead rising above the tower and topped by a cross, soaring to a height of 150 metres – which proved to be its downfall. To quote Stow:

> In the year 1561, the 4th June, betwixt the hours of three and four of the clock in the afternoon, the great spire of the steeple of St Paule's church was fired by lightning, which brake forth (as it seemed) two or three yards beneath the foot of the cross; and from thence it went downward the spire to the battlements, stone-work and bells, so furiously, that within the space of four hours the same steeple, with all the roofs of the church, were consumed, to the great sorrow and perpetual remembrance of the beholders.

Some saw it as a judgement from God for the lack of piety shown by Londoners in their use of the building as a meeting place for many purposes other than religious observance, their very presence attracting into its sacred interior such undesirables as cut-purses and prostitutes plying their trade.

Other than some repairs to the roof and patching up of cracks in the vaulted ceiling and walls, the upper parts of which showed signs of spreading outwards under the lateral loading from the roof, the now spireless structure remained largely neglected until Charles I raised a fund of £100,000 for its restoration. He entrusted the work to Inigo Jones, whose travels in Italy had imbued within him a love of Palladian architecture. After nearly a decade of work, started in 1633, removal of screening and scaffolding revealed nave and transepts clothed in gleaming Portland stone with suppressed buttresses and Italianate windows, but above all a classical Palladian entrance portico at its west end. With much still to be done to the interior and to the tower, the work came to an end with the onset of the Civil War in 1642.

The cathedral suffered grievously under the Parliament. As well as confiscation of the remaining £17,000 in the repair fund, stored materials for use in the repairs and rebuilding were sold, the cathedral plate was melted down, all the woodwork was torn out for use as firewood, stained glass windows were smashed, horses were stabled within the structure and part of it was given over to the incarceration of prisoners. The

Government sold part of the scaffolding, which had been partly supporting the structure, and its removal caused the partial collapse of the stone vaulting and the south transept. Adjacent land was sold to developers and houses, many ramshackle, soon sprang up, right against the cathedral walls.

In 1663, three years after the Restoration of the Monarchy, a Royal Commission set up by Charles II to oversee the repairs to and rebuilding of the structure charged one of its members, John Denham, Surveyor General of the King's Works and sometime poet, to consult with experienced building experts and prepare a report on what measures should be taken. With his conclusion that nothing less than major, and expensive, rebuilding would be needed, the Commission appointed a panel of three architects to plan the work, including Christopher Wren, on whom the responsibility eventually devolved, despite his comparative youth and the fact that he had had only limited architectural experience.

Returning from a visit to Paris in 1666, his imagination seeded by the newly constructed churches he saw there, with their classical features and domed roofs, Wren immediately set about finalising his report on St Paul's to the Commissioners. Unlike Roger Pratt, his fellow architectural expert appointed by the commissioners, who advocated a patchwork of repairs to the old cathedral, Wren recommended that virtually the whole of it should be rebuilt, and with a domed structure that ran counter to the view that an important religious structure such as this should have essentially Gothic features. In a visit to inspect the existing structure with a committee appointed by the commissioners, Wren pointed out many cracks, defects and weaknesses and hung a plumb-line to show the outward lean of the massive Norman pillars caused by the horizontal thrusts from the roof, which Pratt claimed to have been a deliberate construction feature introduced by the Norman builders to give an illusion of added height. The Dean and some other members of the committee apparently, and quite rightly, viewed this with some scepticism.

The committee had made its inspection on 27 August 1666. Six days later, in the early hours of Sunday 2 September, a fire started in a nearby bakery and rapidly spread to the area around St Paul's, which stood directly in its path. It raged for five days and by Friday the whole edifice had been destroyed, masonry shattered by the sheer heat and with melted lead running in streams down Ludgate Hill. There remained no doubt in the mind of William Sancroft, Dean of St Paul's, later to become Archbishop of Canterbury, that the ruins should be cleared away to be completely replaced by a new structure, an opinion fortunately shared by the King. Sancroft also persuaded the King that Christopher Wren should be the architect. Charles II also agreed to contribute £1,000 a year towards expense of rebuilding the cathedral (it is uncertain if he ever did so) and, more importantly, ordered that it should also receive a substantial part of a tax imposed after the Great Fire on coal entering the port of London, to be used for the restoration of the city. This source produced most of the total cost of about £850,000, the balance coming from private donations.

Wren drew up a number of plans before producing one that satisfied all parties. He included, in addition, grandiose plans to replace the narrow winding streets of the city with straight, broad streets radiating from squares, but various vested interests opposed to this ensured that it was not pursued, even though Wren was appointed Surveyor General to the Crown in 1699. Nevertheless, over the next sixteen years he designed fifty-two new churches for the city in a variety of styles. By 1673 Wren had built a wooden model of the new St Paul's, which still survives, of a centrally planned structure dominated by a dome, to be 33 metres in diameter and thus matching that of Hagia Sophia, but some nine to ten metres less than Santa Maria del Fiore and St Peter's. At 156 metres, the length of the new cathedral was 24 metres shorter than its predecessor and the 111 metre height to the top of the cross surmounting the lantern failed by nearly 40 metres to achieve the tip level of the old spire. Despite its total weight of 270,000 tonnes, the building rests on foundations mostly no more than 1.5 metres deep, an exception being the north-east corner of the choir where, sometime in the past, a potter had dug deep to obtain clay and an excavation more than 6 metres deep was needed to reach firm London clay. White limestone from the Portland quarries on the Dorset coast made up the bulk of the masonry in the cathedral, with poorer quality limestone from other sources for use where it had no structural role or had no visual impact.

The external appearance of Wren's dome owed a great deal to St Peter's, although Wren never visited Rome, but the structural form could hardly have been more different. It can be regarded either as an ingenious structure or simply misleading, a charge which has also been made with respect to Wren's use of flying buttresses to convey the horizontal thrusts of the choir vault to the outer walls, which he heightened to hide the flying buttresses from view. The outer ribbed shell of the cupola is not masonry, as in St Peter's, but a dome-shaped timber framework with lead covering. The masonry cupola, made up of some 67,000 tonnes of Portland limestone, is impressive but much flatter seen from the inside. Neither of these structures carries the 700-tonne masonry lantern. The structure provided to carry this is a highly efficient brick cone, between the inner and outer domes, with iron chains at its base to take the hoop tensions set up by the outward thrusts. In the absence of its own weight, or any other load acting on it other than the lantern, the most efficient form for this load carrying structure would be a straight sided cone, but Wren correctly gave it a small convex curvature in elevation to account for self-weight and also to take some of the weight of the timber framework. He may well have discussed this shape with fellow Royal Society member Robert Hooke, in the light of the latter's inverted flexible line analogy to an arch.

The dome has had its problems. Wren rested its 65,000 ton weight primarily on eight piers, each consisting of weak rubble and mortar encased by an outer shell of Portland stone. Under the imposition of the dome, they became out of straight to a sufficient extent that the cathedral was declared a dangerous structure in 1925. Remedial measures put in hand included inserting rustless steel bars into the piers and drilling holes through the outer stone shell to allow liquid cement to be pumped into the interior rubble. Steel bars were also used to brace the inner dome and outer lead-covered dome to help counter cracking observed in the double drum walls supporting the brick cone and lantern. A huge chain with links 4.5 metres long and weighing 30 tons was placed around the outer drum. The work lasted five years and cost about £400,000.

Wren placed the foundation stone with little ceremony in 1675 and 35 years later, in 1710, now 78 years of age and unwilling to make the ascent, he watched as his son, also Christopher, topped out the structure by placing the last stone of the lantern. At that time, the choir had been available for services since 1697 and the north-west chapel since 1699. He had to suffer frustrations and blame for matters outside his control during the

St Paul's, London,
designed by Sir
Christopher Wren.

thirty-five years of construction. His objections to the erection of heavy iron railings around the cathedral were ignored, as was his wish to complete the tops of the cathedral walls with a solid parapet rather than an open balustrade. The commission employed the artist Thornhill to paint the dome despite his disapproval. Serious holdups resulted from a landslide in the Portland quarries, causing delays in delivery of the stones. From 1697 Parliament reduced his stipend from £4 to £2 a week, ostensibly to speed up the work, the commission having accused him of corruption and wilfully delaying the work. In 1718, at the age of 86 and long after completion of the cathedral, he was dismissed from his post as Surveyor General to the Crown. On 25 February 1723, he died in his town house in St James's Street, London at the age of 91.

The eight pillars supporting the dome, and carrying over 7,000 tonnes each, are founded less than 4 metres below churchyard level, or about 1½ metres below crypt level: not on the stiff London clay, but in the water-bearing sands and gravels overlying the clay. In 1831 hundreds of tonnes of sand and silt removed by steam pumps from a pit put down to a depth of nearly 10 metres close to the site of the cathedral caused serious settlement of the piers, ranging from 60 mm to over 150 mm. Cement grout was injected under pressure to stabilise the foundations early in the twentieth century.

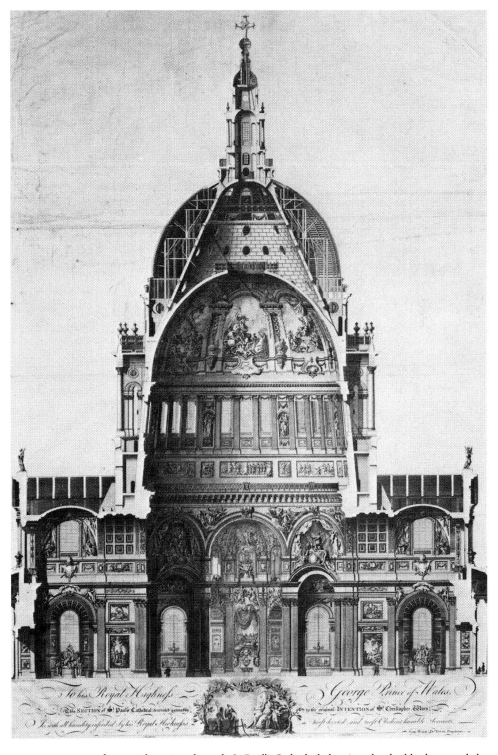

A 1755 engraving of a partial section through St Paul's Cathedral showing the double dome and the supporting brick cone.

Bibliography

Books

Agricola, Georgius, *De Re Metallica*. New York, Dover Publications, 1950.

Akşit, Ilhan, *Istanbul*. Istanbul, Ali Riza Baskan Güel Sanatlar Matbaasi, 1981.

Albert, William, *The Turnpike Road System in England 1663–1840*. Cambridge University Press, 1972.

Alberti, Leon Battista, *The Ten Books of Architecture*. Leoni Edition 1755. New York, Dover Publication, 1986.

An Antiquary (Richard Thomson), *Chronicles of London Bridge*. London, Thomas Tegg, 1839.

Andrews, Francis B., *The Medieval Builder and His Methods*. Wakefield, EP Publishing, 1974.

Al-Hassan, Ahmed Y. and Hill, Donald R., *Islamic Technology*. Cambridge, Cambridge University Press, 1986

Armitage, W. H. G., *A Social History of Engineering*. London, Faber and Faber, 1961.

Barty-King, Hugh, *Water – The Book*. London, Quiller Press, 1992.

Battista, Eugenio, *Brunelleschi*. London, Thames and Hudson, 1981.

Becker, M. Janet, *Rochester Bridge*. London, Constable, 1930.

Blair, John and Ramsey, Nigel, Ed., English Medieval Industries. London, Hambledon and London, 1991.

Bloom, Alan, *The Fens*. London, Robert Hale, 1953.

Bolton, Glorney, *Sir Christopher Wren*. London, Hutchinson, 1956.

Bowyer, Jack, *History of Building*. London, Crosby Lockwood Staples, 1973.

Brangwyn, Frank and Sparrow, Walter Shaw, *A Book of Bridges*. London, John Lane The Bodley Head, 1914.

Briggs, Martin S., *A Short History of the Building Crafts*. Oxford, Clarendon Press, 1925.

Buisseret, David, *Sully*. London, Eyre and Spottiswoode, 1968.

Bullen, F. R., *Notes on the History of Foundation Engineering*. London, The Structural Engineer, December 1961, Vol. XXXIV, No. 12, pp 385–404.

Burstall, Aubrey F., *A History of Mechanical Engineering*. London, Faber and Faber, 1963.

Cook, Martin, *Medieval Bridges*. Princes Risborough, Shire Archaeology, 1998.

Cooper, Michael, *Robert Hooke*. Stroud, Sutton, 2003.

Courtenay, Lynn T., (Ed.), *The Engineering of Medieval Cathedrals*. Aldershot, GB, Ashgate Publishing, 1997.

Cowan, Henry J., *The Masterbuilders*. New York, John Wiley & Sons, 1977.

Crossley, Fred H., *Timber Building in England*. London, Batsford, 1951.

Cresy, Edward, *An Encyclopedia of Civil Engineering, Vol. 1*. London, Thomas Telford, 2001. First published in 1847.

Curzon, George N., *Persia and the Persian Question*. London, Longmans, Green and Co., 1892.

Darby, H. C., *The Draining of the Fens*. Cambridge, Cambridge University Press, 1956.

Darby, H. C., *The Changing Fenland*. Cambridge, Cambridge University Press, 1983.

Davey, Norman, *A History of Building Materials*. London, Phoenix House, 1961.

Dickinson, H. W., *Water Supply of Greater London*. London, Courier Press, 1954.

Duffy, Christopher, *Siege Warfare*. London, Routledge & Kegan Paul, 1979.

Fanelli, Giovanni and Michele, *Brunelleschi's Cupola*. Florence, Mandragora, 2004.

Finch, James Kip, *The Story of Engineering*. New York, Doubleday & Co., 1960.

Finch, James Kip, *Engineering Classics*. Maryland, Cedar Press, 1978.

Fitchen, John, *Building Construction before Mechanization*. Cambridge, Mass., The MIT Press, 1986.

Forbes, R. J., *Studies in Ancient Technology*. Leiden, E. J. Brill, 1965.

Gies, Joseph, *Bridges and Men*. London, Cassell, 1964.

Gies, Frances and Gies, Joseph, *Cathedral, Forge and Waterwheel*. New York, Harper Collins, 1994.

Gille, Bernard, *The Renaissance Engineers*. London, Lund Humphries, 1966.

Gimpel, Jean, *The Medieval Machine*. New York, Holt, Rinehart and Winston, 1976.

Gimpel, Jean, *The Cathedral Builders*. Salisbury, Michael Russell, 1983.

Gough, J. W., *Sir Hugh Myddelton*. Oxford, Clarendon Press, 1964.

Grafton, Anthony, *Leon Battista Alberti*. London, Allen Lane, 2000.

Gregory, J. W., *The Story of the Road*. London. Alexander Maclehose, 1931.

Jackson, Peter, *London Bridge*. London, Cassell, 1971.

Jacobs, David, *Master Builders of the Middle Ages*. London, Cassell, 1970.

Hadfield, Charles, *World Canals*. London, David and Charles, 1986.

Hale, J. R., *Renaissance Fortification*. London, Thames and Hudson, 1977.

Harris, L. E., *Vermuyden and the Fens*. London, Cleaver-Hume Press, 1953.

Hart, Ivor B., *The World of Leonardo Da Vinci*. London, MacDonald, 1961.

Hart-Davis, Adam and Troscianko, Emily, *Henry Winstanley and the Eddystone Lighthouse*. Stroud, Sutton Publishing, 2002.

Harvey, John, *Henry Yevele*. London, Batsford, 1944.

Harvey, John, *The Gothic World*. London, Batsford, 1950.

Harvey, John, *English Mediaeval Architects*. London, Batsford, 1954.

Harvey, John, *The Master Builders*. London, Book Club Associates, 1973.

Hetherington, Paul and Forman, Werner, *Byzantium*. London, Oris Publishing, 1983.

Hewett, Cecil, *English Cathedral Carpentry*. London, Wayland Publishers, 1974.

Heyman, Jacques, *The Stone Skeleton*.Cambridge, Cambridge University Press, 1995.

Hill, Donald, *A History of Engineering in Classical and Medieval Times*. London, Croom Helm, 1984.

Hills, Richard L., *Machines, Mills and Uncountable Costly Necessities*. Norwich, Goose and Son, 1967.

Hindle, Paul, *Medieval Roads and Tracks*. Princes Risborough, Shire Archaeology, 1982.

Hindley, Geoffrey, *A History of Roads*. London, Peter Davies, 1971.

Hislop, Malcolm, *Medieval Masons*. Princes Risborough, Shire Archaeology, 2000.

Home, Gordon, *Old London Bridge*. London, John Lane The Bodley Head, 1931.

Hudson, Kenneth, *Building Materials*. London, Longman, 1972.

Jackson, Gordon, *The History and Archaeology of Ports*. Surrey, World's Work Ltd, 1983.

Jackson, Peter, *London Bridge*. London, Cassell, 1971.

James, John, *The Master Masons of Chartres*. London, West Grinstead Publishing, 1990.

Jardine, Lisa, *On a Grander Scale*. New York, Harper Collins, 2002.

Kaufmann, J. E. and H. W., *The Medieval Fortress*. London, Greenhill Books, 2001.

Kinross, Lord, *Hagia Sophia*. London, Reader's Digest, 1973.

Klemm, Friedrich, *A History of Western Technology*. London, George Allen and Unwin, 1959. Translated by Dorothea Waley Singer.

Knoop, D. and Jones G. P., *The Mediaeval Mason*. Manchester University Press, 3rd ed. 1967.

Lamar, Virginia A., *Travel and Roads in England*. Washington, Folger Shakespeare Library, 1960.

Lang, Jane, *Rebuilding St Paul's*. London, Oxford University Press, 1956.

Lassus, J. B. A., *Album of Villard De Honnecourt*. Paris, Lèonce Laget, Ed., 1976.

Laurenza, Domenico; Taddie, Mario; and Zanon, Edoardo, *Leonard's Machines*. Newton Abbot, David and Charles, 2006.

Maggi, Angelo and Navone, Nicola, (Ed.), *John Soane and the Wooden Bridges of Switzerland*. London. Sir John Soane's Museum, 2003.

Mainstone, Rowland J., *Developments in Structural Form*. London, Allen Lane/Penguin Books, 1975.

Mainstone, Rowland J., *Hagia Sophia*. London, Thames and Hudson, 1988.

Majdalany, Fred, *The Red Rocks of Eddystone*. London, Longmans, 1959.

Maré, Eric de, *The Bridges of Britain*. London, Batsford, 1954.

Maré, Eric de, *Wren's London*. London, The Folio Society, 1975.

Miller, Keith, *St Peter's*. London, Profile Books, 2007.

Milne, Gustav, *Timber Building Techniques in London c. 900-1400*. London and Middlesex Archaeology Society, 1992.

Palladio, Andrea, *The Four Books of Architecture*. New York, Dover Publications, 1965.

Parsons, William Barclay, *Engineers and Engineering in the Renaissance*. Cambridge, Mass., The MIT Press, 1939.

Payne, Robert, *The Canal Builders*. New York, MacMillan, 1959.

Pedretti, Carlo, *Leonardo – Architect*. London, Thames and Hudson, 1986.

Pierce, Patricia, *Old London Bridge*. London, Headline, 2001.

Plumridge, Andrew and Meulenkamp, Wim, *Brickwork*. London, Studio Vista, 1993.

Prager, Frank D. and Scaglia, Gustina, *Brunelleschi – Studies of His Technology and Inventions*. Cambridge, Mass., The MIT Press, 1970.

Procopius, *Buildings Vol. 4* (trans. H. B. Dewing and Glanville Downey). Loeb Classical Library, William Heinemann, London.

Roberts, Gwilym, *Chelsea to Cairo*. London, Thomas Telford, 2006.

Robins, F. W., *The Story of Water Supply*. London, Oxford University Press, 1946

Rolt, L. T. C., *From Sea to Sea*. London, Allen Lane, 1973.

Salzman, L. F., *Building in England*. Oxford, Clarendon Press, 1952.

Saunders, Hilary St George, *Westminster Hall*. London, Michael Joseph, 1951.

Schnitter, Nicholas J., *A History of Dams*. Rotterdam, Balkema, 1994.

Schreiber, Hermann, *The History of Roads*. London, Barrie and Rockliff, 1961.

Skempton, A. W. (Ed.), *John Smeaton FRS*. London, Thomas Telford, 1981.

Smiles, Samuel, *Lives of the Engineers – Early Engineering*. London, John Murray, Popular Edition 1904.

Smith, H. Shirley, *The World's Great Bridges*. London, Phoenix House, 1964.

Smith, Norman, *A History of Dams*. London, Peter Davies, 1971.

Steane, John M., *The Archaeology of Medieval England and Wales*. London, Book Club Associates, 1984.

Steinman, David B. and Watson Sara Ruth, *Bridges and their Builders*. New York, Dover Publications, 1957.

Stow, John, *Survey of London*. London, Dent and Sons, 1956 reprint.

Stratton, Arthur, *Sinan*. London, MacMillan, 1972.

Straub, Hans, *History of Civil Engineering*. London, Leonard Hill, 1960.

Summers, Dorothy, *The Great Level*. London, David and Charles, 1976.

Taverner, John, *Certaine Experiments Concerning Fish and Fruite*. Manchester, Sherratt and Hughes, 1928. Facsimile of 1600 edition.

Tavernor, Robert, *On Alberti and the Art of Building*. London, Yale University Press, 19098.

Toy, Sidney, *A History of Fortification*. London, William Heinemann, 1955.

Vasari, Giorgio, *Artists of the Renaissance*. London, Allen Lane, 1978 ed.

Warner, Philip, *The Medieval Castle*. London, Book Club Associates, 1973.

Watts, Martin, *Water and Wind Power*. Princes Risborough, Shire Publications, 2000.

Welch, Charles, *History of the Tower Bridge*. London, Smith, Elder and Co., 1894.

White, Lynn Jr., *Medieval Technology and Social Change*. Oxford University Press, 1962.

Whitney, Charles S., *Bridges – Their Art, Science and Evolution*. New York, Greenwich House, 1983.

Wilkinson, T. W., *From Track to By-Pass*. London, Methuen & Co., 1934.

Willan, T. S., *River Navigation in England 1600–1750*. London, Frank Cass and Co., 1964.

Wright, Esmond (Ed.), *The Medieval and Renaissance World*. London, Hamlyn, 1979.

Zammattio, Carlo; Marinoni, Augusto; and Brizio, Anna Maria, *Leonardo the Scientist*. London, Hutchinson, 1981.

Papers, Articles and Contributions

Binda, L., et al, 'The Collapse of the Civic Tower of Pavia: A Survey of the Materials and Structure.' *Masonry International*, 1992, Vol. 6, No. 1, pp 11–20.

Briggs, Martin S., 'Building Construction'. *A History of Technology*, Oxford Clarendon Press, 1956, Vol. 2, pp 397–448.

Briggs, Martin S, 'Building Construction'. *A History of Technology*, Oxford Clarendon Press, 1957, Vol. 3, pp 245–268.

Bullen, F. R., 'Notes on the History of Foundation Engineering'. *Jour. Inst. Structural Engineers*, 1961, Vol. XXXIX, No. 12, pp 385–404.

Burland, John; Jamiolkowski, Michele and Viggiani, Carlo, 'Preserving Pisa's Treasure'. *Civil Engineering*, 2002, March, pp 42–9.

Burland, J. B., 'Stabilising the Leaning Tower of Pisa: the Evolution of Geotechnical Solutions'. *Transactions of the Newcomen Society*, 2008, Vol. 78, No. 2, pp 173–205.

Choay, Françoise, 'Alberti and Vitruvius'. *Architectural Design*, Vol. 49, No. 5–6, pp 26–9.

Evans, F. T., 'Monastic Multinationals: the Cistercians and other Monks as Engineers'. *Transactions of the Newcomen Society*, 1996–7, Vol. 68, pp 1–28.

Fairclough, K. R., 'The Waltham Pound Lock'. *History of Technology*, Mansell, London, 4th Annual Volume, 1979, pp 31–44.

Goodchild, R. G. and Forbes, R. J., 'Roads and Land Travel, with a Section on Harbours, Docks and Lighthouses'. *A History of Technology*, Oxford Clarendon Press, 1956, Vol. 2, pp 493–536.

Grayson, Cecil, 'Leon Battista Alberti, Architect'. *Architectural Design*, Vol. 49, No. 5–6, pp 7–17.

Hall, A. R., 'Military Technology'. *A History of Technology*, Oxford Clarendon Press, 1957, Vol. 3, pp 347–76.

Heyman, Jacques, 'The Stone Skeleton'. *Int. Jour. Solids Structures*, 1966, Vol. 2, pp 249–79.

Hamilton, S. B., 'Bridges'. *A History of Technology*, Oxford Clarendon Press, 1957, Vol. 3, pp 417–37.

Reynolds, Terry S., 'The Phoenix and its Demons: Waterpower in the Past Millenium'. *Transactions of the Newcomen Society*, 2006, Vol. 76, No. 2, pp 153–74.

Skempton, A. W., 'Canals and River Navigations before 1750'. *A History of Technology*, Oxford Clarendon Press, 1957, Vol. 3, pp 438–70.

Skempton, A. W., 'Engineering on the English River Navigations'. In *Canals – A New Look*, edited by Mark Baldwin and Anthony Burton, London, Phillimore, 1984 pp 23–44.

Acknowledgements

My many thanks to John Burland for looking through my coverage of the Pisa Tower and for his suggested amendments. As a member of the Sixteenth Commission, John played a leading role in stabilising the tower. My thanks, too, to Campbell McCutcheon, Louis Archard and their colleagues at Amberley Publishing for making the book a reality.

Picture Sources

14, 15, 69t: Bibliothéque Nationale, Paris. 28b, 67, 105, 106: (Codex Atlanticus) Ambrosiana Library, Milan. 48: J. Bowles. 53: Pom² (Wikipedia). 63: Lorini, *Delle fortificationi* (1596). 65: Al Jasari (thirteenth century), Süleymaniye Library, Istanbul. 69b: Bertrand Gille. 77: Samuel Smiles, *Lives of the Engineers*. 79: John Bate, *Mysteries of Nature and Art*, 1635 wood-cut. 93: Darby, *The Changing Fenland*. Cambridge University Press. 94: G. Fowler, 1932. 103, 114b; (Skempton): *A History of Technology, Vol. 3*, Oxford University Press. 112: Adapted from L. T. C. Rolt, *From Sea to Sea* (Rolt Estate). 114t: Boerkevitz (Wikipedia). 123: Luttrell Psalter, fourteenth century. 124: Weigel, *Book of the Estates*, 1698. 131: Palladio, *The Four Books of Architecture*, 1570. 135: Bibliothèque des Arts Décoratifs, Paris. 145: Engraving by John Norden, *c.* 1600. 147: Drawing by Gordon Home, *c.* 1930, British Library. 156b: Roland Mainstone. 170: Herbert Ortner (Wikipedia). 171: Sidney Toy. 172: Xvlun (Wikipedia). 199t: John Muller, 1746. 179b: Belidor, *La Science des Ingenieurs*. 184, 185: Engravings by Roberts, 1762, British Museum. 190: Gryffindor (Wikipedia). 201: John Burland. 205: Mossot (Wikipedia). 210: Library of Congress. 214: Adapted from J. RIBA, Vol. 5, 1907–8. 216: Viollet-le-Duc, *Dictionnaire Raisonné de l'Architecture du XIe `a XVIe Siecle*, 1848–58. 218: Library of Congress. 219: Batsford/Salisbury Cathedral. 235: Prager and Scaglia, 1970 (drawn by G. Rich, 1969), MIT Press. 241: Library of Congress. 247: Engraving by unknown artist, 1755, Guildhall Library.

Other images from the Author's collection.

Index

Principal Names

General Index